D0500451

BITWISE

BITWISE

A Life in Code

DAVID AUERBACH

PANTHEON BOOKS, NEW YORK

Published in the United States by Pantheon Books, a division
of Penguin Random House LLC, New York, and distributed in Canada
by Random House of Canada, a division
of Penguin Random House Canada Limited, Toronto.

Pantheon Books and colophon are registered trademarks of Penguin Random House LLC.

Grateful acknowledgment is made to Schocken Books, a division of
Penguin Random House LLC, for permission to reprint an excerpt
of "In the Penal Colony" from *The Completed Stories* by Franz Kafka,
edited by Nahum N. Glatzer, copyright © 1946, 1947, 1948, 1954, 1958, 1971
by Penguin Random House LLC. Reprinted by permission of Schocken Books,
a division of Penguin Random House LLC. All rights reserved.

Some material in the chapters "Logo and Love" and "Chat Wars"
first appeared, in a different form, in *Slate* and *n + 1*.

Page 291 constitutes an extension of this copyright page.

Library of Congress Cataloging-in-Publication Data
Name: Auerbach, David (David B.), author.
Title: Bitwise : a life in code / David Auerbach.
Description: First edition. New York : Pantheon Books, 2018.
Includes bibliographical references and index.
Identifiers: LCCN 2017055983. ISBN 9781101871294 (hardcover : alk. paper).
ISBN 9781101871300 (ebook).
Subjects: LCSH: Computer science—Philosophy. Computer science—Social aspects.
Auerbach, David (David B.)—Philosophy. Computer scientists—United States—Biography.
Classification: LCC QA76.167 .A84 2018 | DDC 004—dc23 |
LC record available at lccn.loc.gov/2017055983

www.pantheonbooks.com

Jacket design by Tyler Comrie
Book design by Maggie Hinders

Printed in the United States of America

First Edition

2 4 6 8 9 7 5 3 1

For Nina, Eleanor, and Iris

CONTENTS

BITWISE

INTRODUCTION

> Thoughtfulness means: not everything is as obvious as it used to be.
>
> —HANS BLUMENBERG

COMPUTERS always offered me a world that made sense. As a child, I sought refuge in computers as a safe, contemplative realm far from the world. People confused me. Computers were precise and comprehensible. On the one hand, the underspecified and elusive world of human beings; on the other, the regimented world of code.

I had tried to make sense of the real world, but couldn't. Many programmers can. They navigate relationships, research politics, and engage with works of art as analytically and surgically as they do code. But I could not determine the algorithms that ran the human world. Programming computers from a young age taught me to organize thoughts, break down problems, and build systems. But I couldn't find any algorithms sufficient to capture the complexities of human psychology and sociology.

Computer algorithms are sets of exact instructions. Imagine describing how to perform a task precisely, whether it's cooking or dancing or assembling furniture, and you'll quickly realize how much is left implicit and how many details we all take for granted without giving it a second thought. Computers don't possess that knowledge, yet computer systems today have evolved imperfect pictures of ourselves and our world. There is a gap between those pictures and reality. The smaller the gap, the more useful computers become to us. A self-driving car that can only distinguish between empty space and solid objects operates using a primitive image of the world. A car that can distinguish between human and nonhuman objects possesses a more sophisticated picture, which makes it better able to avoid deadly errors. As the gap closes, we can better trust computers to *know* our world. Computers can even trick us into thinking the gap is smaller than it really is. This book is about that gap, how it is closing, and how *we* are changing as it closes. Computers mark the latest stage of the industrial revolution, the next relocation of our experience from the natural world to an artificial and man-made one. This computed world is as different from the "real" world as the factory town is from the rural landscape.

Above all, this book is the story of my own attempt to close that gap. I was born into a world where the personal computer did not yet exist. By the time I was old enough to program, it did, and I embraced technology. In college, I gained access to the internet and the nascent "World Wide Web," back in the days when AOL was better known than the internet itself. I studied literature, philosophy, and computer science, but only the latter field offered a secure future. So after college I took a job as a software engineer at Microsoft before moving to Google's then-tiny New York office. I took graduate classes in literature and philosophy on the side, and I continued to write, even as the internet ballooned and our lives gradually transitioned to being online all the time. As a coder and a writer, I always kept a foot in each world. For years, I did not understand how they could possibly converge. But neither made sense in isolation. I studied the humanities to understand logic and programming, and I studied the sciences to understand language and literature.

A "bitwise operator" is a computer instruction that operates on a sequence of bits (a sequence of 1s and 0s, "bit" being short for "binary digit"), manipulating the individual bits of data rather than whatever those bits might represent (which could be anything). To look at something bitwise is to say, "I don't care what it means, just crunch the data." But I also think of it as signifying an understanding of the hidden layers of data structures and algorithms beneath the surface of the worldly data that computers store. It's not enough to be worldwise if computers are representing the world. We must be bitwise as well—and be able to translate our ideas between the two realms.

This book traces an outward path—outward from myself and my own history, to the social realm of human psychology, and then to human populations and their digital lives. Computers and the internet have flattened our local, regional, and global communities. Technology shapes our politics: in my lifetime, we have gone from Ronald Reagan, the movie star president, to Donald Trump, the tweeting president. We are bombarded with worldwide news that informs our daily lives. We form virtual groups with people halfway around the world, and these groups coordinate and act in real time. Our mechanisms of reason and emotion cannot process all this information in a systematic and rational way. We evolved as mostly nomadic creatures living in small communities, not urban-dwelling residents connected in a loose but extensive mesh to every other being on the planet. It's nothing short of astounding that the human mind copes with this drastic change in living. But we don't think quite right for our world today, and we are attempting to off-load that work to computers, to mixed results.

Computers paradoxically both mitigate and amplify our own limitations. They give us the tools to gain a greater perspective on the world. Yet if we feed them our prejudices, computers will happily recite those prejudices back to us in quantitative and apparently objective form. Computers can't know us—not yet, anyway—but we think they do. We see ourselves differently in their reflections.

We are also, in philosopher Hans Blumenberg's term, "creatures of deficiency." We are cursed to be aware of our poverty of understanding and the gaps between our constructions of the world and the world itself, but we can learn to constrain and quantify our lack of understanding. Computers may either help us understand the gaps in our knowledge of the world and ourselves, or they may exacerbate those gaps so thoroughly that we forget that they are even there. Today they do both.

PART I

1

LOGO AND LOVE

The Turtle

> I found particular pleasure in such systems as the differential gear. . . . I fell in love with the gears.
>
> —SEYMOUR PAPERT

WE ARE DRIVEN TO DISCOVER how things work, but I was often disappointed to find out that one thing or another didn't work more neatly. The television, the automobile, and the human body seemed like they could be more organized, more elegant. Computers, however, did not disappoint me.

Like so many software engineers, I was a shy and awkward child, and I understood computers long before I understood people. The precision, clarity, and reliability that computers promised, particularly in the 1980s when they were so much simpler than they are today, provided a refuge for many children who did not easily integrate into the social fabric of their peers. But a computer was not merely something that I could play with; it was something I could program and control, and with which I could create a new world. Computers are now moving toward virtual reality and photorealistic games, but back then computers displayed only a screen of text and primitive monochrome graphics, which were nonetheless enough to support something that remains more fundamentally powerful than the sharpest graphics: code.

My first computer language was Logo, a graphical language developed in 1967 by Wally Feurzeig, Seymour Papert, and Cynthia Solomon and intended as an educational tool. I learned it at a computer class for kids at our local rec center in the suburbs of Los Angeles when I was seven. Armed with Logo, I could write instructions (in the form of a program) for a triangular "turtle" on the screen, which would then draw lines and shapes based on those instructions. The screen was monochrome, green text and lines on a black background.

The first "program" I wrote was a single line of code: drawing a square.

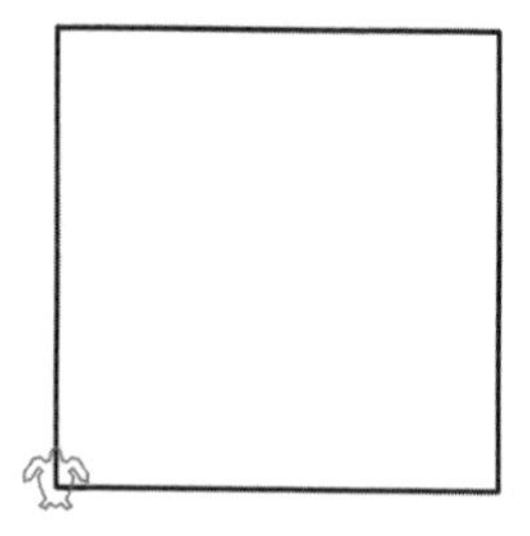

repeat 4 [forward 50 right 90]

That is, go forward 50 pixels, turn right by 90 degrees, and then repeat those two steps a total of four times. At the end of it, the turtle would be back where it started, having drawn out a square. By changing the angle and the number of repeats, I could draw a variety of polygons. A triangle:

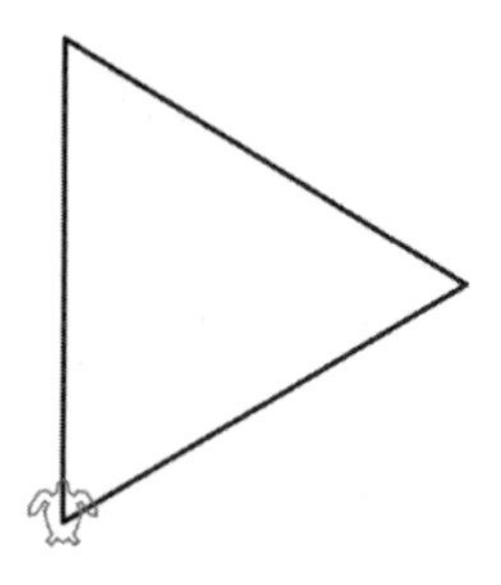

repeat 3 [forward 50 right 120]

An octagon:

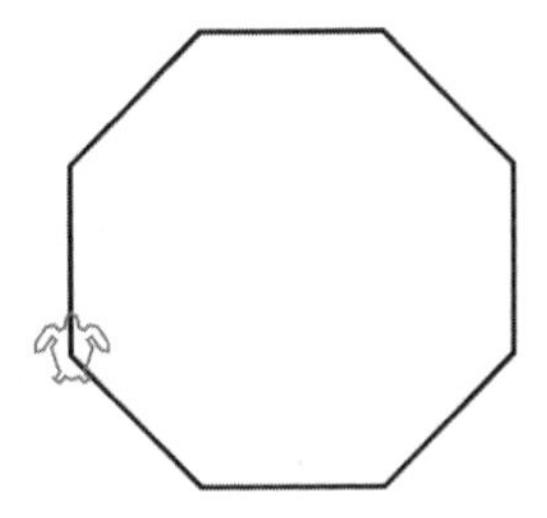

repeat 8 [forward 50 right 45]

A pentagram:

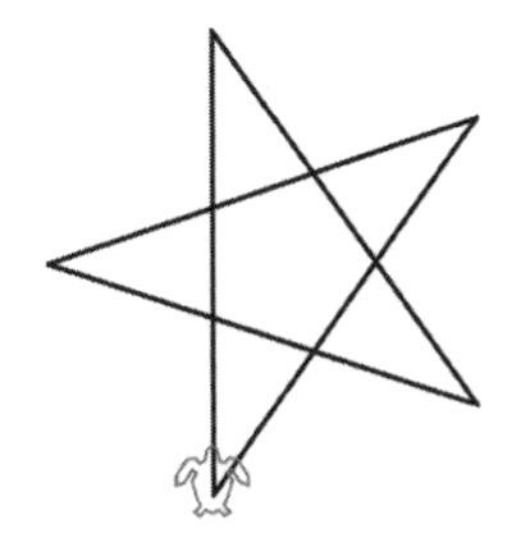

repeat 5 [forward 50 right 144]

I could not draw a pentagram by hand, at least not well. The turtle drew it perfectly. The 144-degree angle felt like secret knowledge to me. I hadn't realized that the program did not need to be any more complex than that for a square or an octagon. Sometimes I boosted the number of repeats so that the turtle would continue to zip along the pentagram's lines like a bullet train.

These single-line programs are all algorithms. The word "algorithm" is a derivation of the name of ninth-century Persian mathematician Muhammad ibn Mūsā al-Khwārizmī. An algorithm is, informally speaking, the set of rules or instructions specifying the path from a specified problem ("Draw a pentagram with sides of length 50") to the solution to that problem (the visual display of the pentagram itself). Algorithms can become increasingly general, specified with variables rather than constants ("Draw a polygon with *n* sides of length *m*").

Algorithms hooked me. My own experience suggests that some

people's brains are more tuned in to this way of thinking, just as some people are more attuned to mathematics or languages. I am not a visual or a verbal person: I was rejected from kindergarten because I couldn't draw. But these kinds of assemblages of instructions made intuitive sense, and I thought they were beautiful. Instead of just having the thing itself, I had the recipe for the thing and, moreover, could make the recipe increasingly general so that reams of problems could be solved by twiddling the dials on a single recipe. That, in essence, is computer programming.

Simple algorithms can produce beautifully complex results. Here is a Logo program of half a dozen lines, **sierpinskiTriangle**, which draws a fractal triangle.

```
to sierpinskiTriangle :length :depth
     if :depth < 1 [ stop ]
     repeat 3 [
          sierpinskiTriangle :length/2 :depth-1
          forward :length
          right 120
     ]
end
```

Invoking the program with the command **sierpinskiTriangle 500 7** will cause the turtle to draw the following graphic:

You can get this fractal pattern out of six lines of code because **sierpinskiTriangle** is doing one thing over and over again: drawing a triangle made out of three triangles. But every time it draws one of those triangles, it first draws three smaller triangles *inside* that triangle—in other words, it does the same thing, just smaller. So the code calls itself, in a process called *recursion.*

Here is another example of recursion, a program to draw a tree:

```
to tree :level :size :scale :angle
     if :level > 0 [
          fd :size
          lt :angle
          tree :level - 2 :size * :scale * :scale :scale :angle
          rt :angle
          rt :angle
          tree :level - 1 :size * :scale :scale :angle
          lt :angle
          bk :size
     ]
end
```

Invoking this program with **tree 18 100 .9 20** produces this graphic:

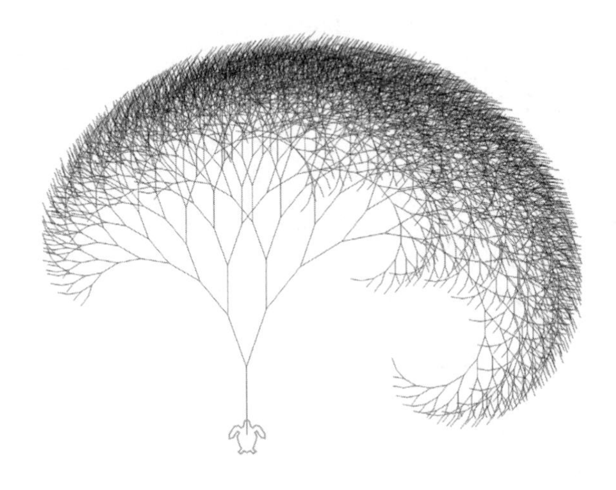

This amazed me. It seemed impossible. How could a dozen lines of code produce such a beautiful and complex pattern? How had I instructed this computer to draw more capably and more beautifully than my hand could? I wanted to understand how such a great effect could stem from such a small set of instructions, and I wanted to author the programs that created such effects. My confusion led to my desire to understand. My wonder led to my desire to create.

Many people find their calling in a moment of sheer awe. The awe stems from not just the beauty and elegance but the sheer seeming *impossibility* of a past creation or discovery. For a writer, this could occur on reading a particular line of Shakespeare or Zhuangzi. For a mathematician, it may be found when studying the proof of the irrationality of the square root of 2 or the supremely elegant unity of Euler's equation, which joins five fundamental mathematical constants through addition, multiplication, and exponentiation:

$$e^{i\pi} + 1 = 0$$

I was impressed and perplexed by this equation when I first saw it. The relation of the constants isn't obvious. I found it beautiful, yet it did not impel me to devour books of mathematics. I could appreciate the elegance of Euler's equation without wishing to dissolve my identity in the world of mathematics. Not so with the world of computers.

Plato believed that the core impulse to philosophizing lies in *aporia*, the point at which, in struggling to understand a phenomenon or answer a question, we come up against a seemingly irresolvable contradiction. The force of this contradiction can make us reassess the totality of what we thought we knew and reformulate it in a revolutionary way—for example, by saying, as Copernicus did in 1543, "Yet at rest in the middle of all things is the Sun."

Elsewhere, in the *Theaetetus,* Plato writes that philosophy begins in wonder (*thaumazein*), the awe-inspiring excitement that I felt on seeing the turtle-drawn tree. Aristotle, Plato's stolid successor, played down *aporia*. Perhaps this helped him to generate answers far more readily than Plato did. Aristotle produced systems of earthly and celestial motion, attempts at basic biology, and classifications of the various peoples and humors of the world.* Plato's works, instead, tend to dwell more on how easily our minds are misled, and ask how we can be certain of anything. We care and work hard to understand and master a discipline through a combination of wonder and confusion. In his 1938 book *Experience and Prediction*, the philosopher of science Hans Reichenbach described the human condition as one not just of profound ignorance, but also illusion:

> We walk through the world as the spectator walks through a great factory: he does not see the details of machines and working operations, or the comprehensive connections between the different departments which determine the working processes on a large scale. . . . We see the polished surface of our table as a smooth plane; but we know that it is a network of atoms with interstices much larger than the mass particles, and the microscope already shows not the atoms but the fact that the apparent smoothness is not better than the "smoothness" of the peel of a shriveled apple. We see the iron stove before us as a model of rigidity, solidity, immovability; but we know that its particles perform a violent dance, and that it resembles a swarm of dancing gnats more than the picture of solidity we attribute to it. We see the moon as a silvery disk in the celestial vault, but we know it is an enormous ball suspended in open space. We hear the voice coming from the mouth of a singing girl as a soft and continuous tone, but we know that this sound is composed of hundreds of impacts a second bombarding our ears like a machine gun. . . . We do not see the things, not even the concreta, as they are but in a distorted form; we see a substitute world—not the world as it is, objectively speaking.

This "substitute world" that we see is, in short, a lie. Our brains take sense data and inaccurately analogize it into forms that are already

* Aristotle's answers, particularly in the natural sciences, were usually wrong. His *Physics* is an ambitious compendium of mistakes and misapprehensions, drawn from intuition rather than science. But Aristotle's genius lay in claiming new territory—not in mapping it accurately. If he had a fault, it was being far too easily satisfied. Perhaps if he had entertained more doubts about accuracy, he wouldn't have treated as many subjects as he did.

familiar to us. But as children growing up, this substitute world works quite well. It is manageable and legible to us, since we engage with the world in a functional and effective fashion. The world as it is only grants flashes of strangeness to a child to suggest that reality might be quite different. What really goes on inside our bodies? How did this world come to be? What is death? These questions don't often present themselves, because we know to be productive with our time rather than diving into what David Hume called the "deepest darkness" of paralyzing skepticism. Yet much joy and satisfaction can be found in chasing after the secrets and puzzles of the world. I felt that joy first with computers. In them I found a world strictly divided between the program and the output. The instructions and the execution. The simple code and the beautiful tree.

I remain a terrible artist, barely able to draw a human figure. But I fell in love with the concepts of algorithmic programming: instructions, branches, variables, functions. I saw how a program could generate the simulated world of the output. Recursion was too tricky for my seven-year-old self to wrap my head around, but I wanted badly to understand it, and I was convinced that I could. My wonder at the power of programming, the ability to create merely through simple lines of text and numbers, drove me.

The complexity of life is all around us, but we grow numb to what we see of it, even while so much lies outside our immediate experience: microworlds of cells, atoms, particles, as well as the macrocosmos of our universe containing far more galaxies than the Earth has people (approximately two trillion galaxies by NASA's 2016 estimate). Programming abstracted away the uncertainty of the world and laid its principles out before me. Notions of elegance and beauty drive programmers just as much as they do mathematicians and poets. What mattered is that I *felt* the jump from the programmatically simple to the aesthetically complex.

On a computer, that jump is clean, elegant, and definitive. One popular philosophical fable concerns the myth that the Earth is supported on the shell of a gigantic turtle.* "What is supporting the turtle?" asks

* The turtle myth occurs most prominently in Northeastern Native American mythology, but the source of the philosophical fable, made most famous by Stephen Hawking in *A Brief History of Time*, is less certain. The legal scholar Roger Cramton attributes it to an anonymous commentator on William James, who told the fable with rocks instead of turtles.

the philosopher. It's "turtles all the way down," comes the reply. There is no final answer available to us, only more questions. Programming offered a stopping point with its artificial world, a final answer. In Logo, there was just the turtle, just the one.

The first turtle I worked with was a simple triangle, not the waddling shape you see in the pictures above. Later, the program Logo Writer made an appearance at my school. It was a frillier version of Logo, which replaced the triangle with a turtle shape closer to what I've used here. I disliked the turtle-shaped turtle (and I still do). Logo Writer added bells and whistles, but the representation of the turtle *as a turtle* had no functional impact whatsoever on the workings of Logo. It was a superfluous cosmetic change that drew attention away from what was truly remarkable about Logo: the relationship between the program and its execution. The turtle, whether triangle-shaped or turtle-shaped, was already abstracted away in my mind, just a point to designate where drawing would next originate.

Even as I coded on Logo Writer, that tree still puzzled me. I could not understand the concept behind recursion, the powerful technique that allowed the `tree` program to draw such a complicated pattern with so few lines of code. I wouldn't figure it out until my teens, when I would also learn what a powerful role it played in all of computer science and indeed in conceptual thinking in general. Recursion, in a nutshell, is use of a single piece of code to tackle a problem by breaking it down into subproblems *of the same form*—like drawing a branch of a tree that is itself a smaller tree. It is envisioning the world as an ornate yet fundamentally elegant fractal. Recursion reflects the efficient, parsimonious instinct of computer programming, which is to get a lot out of a little.

The Assembly

> Why would you want more than machine language?
>
> —JOHN VON NEUMANN

Programming isn't a wholly abstract exercise. Programming requires hardware, which only became available to the average home in the

1980s, and it had its own subculture too. In the pre-internet days, there was a secret lore surrounding computers, and much of it revolved around the Apple II.

My first computer was an Apple IIe. It was, by far, the most popular home computer of its age, thanks not only to Apple's partnerships with educators but also to Apple's focus on making a *general-purpose* computer for consumers and hobbyists. Other consumer-oriented computers, like the Commodore 64 or the Atari 400, were dedicated simply to running the primitive software of the time. Apple's computers used Chuck Peddle's ubiquitous (and *cheap*) 6502 processor and sat somewhere in between those casual machines and professional PCs like IBMs. This owes primarily to Apple creator Steve Wozniak's background in the hobbyist and computer club community and his dedication to building a computer that could be both accessible and powerful.* Wozniak was not trained in academia or research labs. He came out of vaguely countercultural groups who got into PCs with the same fervency that others get into coin collecting, cars, or Dungeons & Dragons.†

There was a totality to the Apple IIe that no longer exists on computers today, or even mobile devices. It offered the sense of being *close* to the fundamental machinery of the system. The Apple IIe did not have a hard drive. Turn it on without a floppy in the drive and you'd just see "Apple][" frozen at the top of the monitor. I had to boot a floppy disk containing Apple DOS, the disk operating system, where I could program in Applesoft BASIC, as did many others around that time.

I remember the first programs I tinkered with on the Apple IIe. There was *Lemonade Stand*, a multiplayer accounting game originally created by Bob Jamison back in 1973, then ported to the Apple IIe in 1979 by Charlie Kellner. Playing it, I set prices and budgeted for advertising, depending on the weather. If prices were too high, people wouldn't

* The ranks of hobbyists notably included literary critic and polymath Hugh Kenner, who, while writing eclectic studies like *The Poetry of Ezra Pound*, *The Stoic Comedians: Flaubert, Joyce, and Beckett*, *Geodesic Math and How to Use It*, and *Chuck Jones: A Flurry of Drawings*, found time to author the *Heath/Zenith Z-100 User's Guide* (1984) and convince his friend William F. Buckley Jr. to switch to a word processing program in 1982.

† At the time, I and many other computer enthusiasts thought of Wozniak as *the* visionary star of Apple. Steve Jobs was seen as little more than Wozniak's handler. Back then, Wozniak was the more approachable face for a company like Apple than the comparatively uptight and slick Jobs. How times change.

buy your lemonade. If prices were too low, you wouldn't make a profit. After a few days of play, your mother stops giving you free sugar; a few days after that, the price of lemonade mix goes up. If it rained, everything was destroyed for that day and you took a total loss. If construction crews were present on the street, they would pay *any* price for your lemonade. I changed the code so *everyone* would pay whatever price you set. I made a killing because I could change the rules. Then I changed the code so that for the second player only, people would never buy lemonade at any price, and I asked my mother to play it against me. I won. She was baffled, then simultaneously impressed and annoyed (a reaction that is every child's dream) when I told her I'd changed the code.

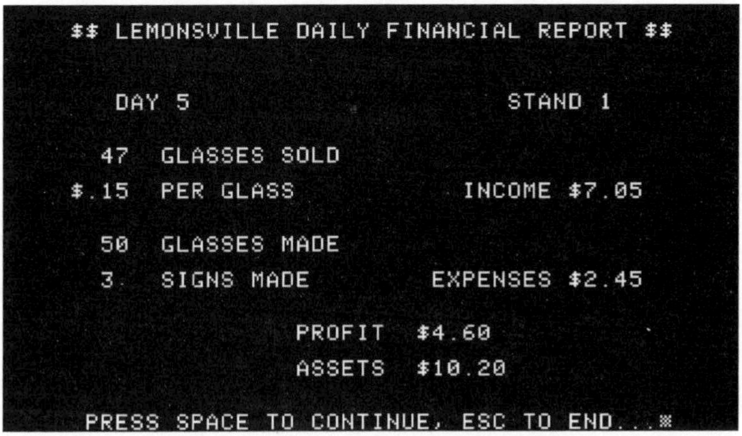

```
$$ LEMONSVILLE DAILY FINANCIAL REPORT $$

   DAY 5                      STAND 1

  47  GLASSES SOLD
$.15  PER GLASS            INCOME $7.05

  50  GLASSES MADE
  3   SIGNS MADE         EXPENSES $2.45

            PROFIT  $4.60
            ASSETS  $10.20

 PRESS SPACE TO CONTINUE, ESC TO END...
```

Profiting from a heavy markup in *Lemonade Stand*

BASIC was a less elegant language next to Logo. But it was *native* to the Apple IIe. With a few mysterious commands named PEEK, POKE, and CALL, I could tinker directly with the guts of the Apple IIe. These commands let you access the physical Random Access Memory (RAM) of the machine, the immediate, transient short-term storage of the computer. **PEEK(49200)** would make the speaker *click*, a thrilling sound when the single loud beep was the only sound easily available to a BASIC programmer. **POKE(49384,0)** would start the disk drive motor spinning, good for scaring someone into thinking their disk was being formatted. Other PEEKs and POKEs allowed for manipulation of text to make characters disappear and reappear and move around—things that weren't easy to do in BASIC proper. You could also crash and reboot your machine, which was otherwise nearly impossible in BASIC. POKEs and CALLs were powerful stuff. PEEK was (mostly) safe.

These numbers were arbitrary to outsiders, and they were not particularly publicized. Before the internet, programmers had to learn this kind of esoteric knowledge haphazardly from books, magazines, and other enthusiasts. There was a thrill of discovery that can't be recreated now that most information can be found with a simple web search.* I would find a particular piece of Apple lore, then think about it until the next time I got on the computer to try it out. I discovered much, as many did then, through the charts produced by Bert Kersey's Beagle Bros software company. There is a certain set of people, myself included, for whom this chart will inspire an overpowering nostalgia.

Part of this nostalgia owes to Kersey's signature design and clip art, which distinguished Beagle Bros from other vendors. Part of it owes to the sheer intrigue around the secret details contained on the chart: this was hidden knowledge! Kersey captured the mystique:

> Pokes are often used to write machine-language routines that may be activated with the CALL command—the possibilities are infinite.

Even novelties were fascinating because they revealed unsuspected capabilities of the Apple IIe. The program, which appeared in a Beagle Bros catalogue, was pretty much impossible to parse if read.

```
1 HOME: LIST: BUZZ=49200
2 A$="!/-"+CHR$(92): FOR A=1 TO 48: B=PEEK(BUZZ):FOR C=1 TO
    A: NEXT: X$=MID$ (A$,A-INT(A/4)*4+1,1): VTAB 3: HTAB
    10: PRINT X$X$X$: NEXT: GOTO 2
```

If typed in and executed, it would print itself and then make a varispeed buzz as the characters in the first line appeared to spin around

* Musician Julian Cope describes the mystique of the Velvet Underground's bootleg "Foggy Notion" record: "With the accessibility of music via Napster and gemm.com, it's difficult for all you young'uns to understand how mysterious unreleased and bootlegged material was back then."

Peeks, Pokes and Pointers

Apple® Zero-Page

DECIMAL		HEX
32	**Text Window Left-Edge** (0-39 / normal is 0) Example: **POKE 32, X** freezes the left X columns of text. Warning: Don't let PEEK(32)+PEEK(33) exceed the screen width.	$20
33	**Text Window Width** (1-40 or 1-80 / normal is 40 or 80) Note: **POKE 33,33** scrunches listings to remove extra spaces.	$21
34	**Text Window Top-Edge** (0-23 / normal is 0)	$22
35	**Text Window Bottom** (1-24 / normal is 24)	$23
36	**Horizontal Cursor-Position** (0-39) Examples: If **PEEK(36)=X**, then the cursor is in column X+1. **POKE 36,X** puts the cursor in column X+1 (useful with 80-columns, for positioning the cursor beyond the 40-column limit of HTAB). Note: **POKE 1403,X** works similarly—and more predictably.	$24
37	**Vertical Cursor-Position** (0-23) Examples: If **PEEK(37)=Y**, then the cursor is on text line Y+1.	$25
43	**Boot Slot★16** (after boot)	$2B
44	**Lo-Res Line End-Point**	$2C
48	**Lo-Res COLOR★17**	$30
50	**Text Output Format** POKE 50, **63**=INVERSE, POKE 50, **255**=NORMAL, POKE 50, **127**=FLASH (for ASCII 64-95).	$32
51	**Prompt-Character** Note: **POKE 51,0: GOTO line#** will prevent a false "Not Direct Command" message caused by an *immediate* **GOTO line#** command.	$33
78-79	**Random-Number Field**	$4E.4F
103-104	**Start of Applesoft Program** To Load a program at a non-standard location LOC— POKE LOC-1, 0: POKE 103, LOC-INT(LOC/256)*256: POKE 104, INT(LOC/256) *Then* LOAD PROGRAM Note: **FP** (DOS 3.3 only) sets start-of-program to normal 2049 ($801).	$67.68
105-106	**LOMEM** Note: LOMEM is the Start of Variable-Space, equivalent to End-of-Program (approx.) unless changed with the **LOMEM:** command.	$69.6A
107-108	**Start of Array-Space**	$6B.6C
109-110	**End of Array-Space**	$6D.6E
111-112	**Start of String-Storage**	$6F.70
115-116	**HIMEM** Note: HIMEM-1 is the highest address available for use by an Applesoft program. May be changed with the **HIMEM:** command.	$73.74
117-118	**Line-Number Being Executed**	$75.76
119-120	**Line-No. Where Program Stopped**	$77.78
121-122	**Address of Line Executing**	$79.7A
123-124	**Current DATA Line-Number**	$7B.7C
125-126	**Next DATA Address**	$7D.7E
127-128	**INPUT or DATA Address**	$7F.80
129-130	**Last-Used Variable Name**	$81.82
131-132	**Last-Used-Variable Address**	$83.84
175-176	**End of Applesoft Program**	$AF.B0
214	**RUN Flag** Example: **POKE 214, 255** makes *any* command RUN a program.	$D6
216	**ONERR Flag** Example: **POKE 216, 0** cancels the ONERR function.	$D8
218-219	**Line-Number of ONERR Error**	$DA.DB
220-221	**ONERR Error Address**	$DC.DD
222	**ONERR Error Code**	$DE

DOS 3.3 and PRODOS	APPLESOFT
1: Language Not Available[1]	0: ?Next Without For
2 or 3[1]: Range Error	16: ?Syntax Error (FP)
3: No Device Connected[2]	22: ?Return Without Gosub
4: Write-Protected	42: ?Out of Data
5: End of Data	53: ?Illegal Quantity
6: File[1] or Path[2] Not Found	69: ?Overflow
7: Volume Mismatch[1]	77: ?Out of Memory
8: I/O Error	90: ?Undef'd Statement
9: Disk Full	107: ?Bad Subscript
10: File Locked	120: ?Redim'd Array
11: Syntax Error[1] or Invalid Option[2]	133: ?Division by Zero
12: No Buffers Available	163: ?Type Mismatch
13: File Type Mismatch	176: ?String Too Long
14: Program Too Large	191: ?Formula Too Complex
15: Not Direct Command	224: ?Undef'd Function
17: Directory Full[2]	254: ?Re-Enter
18: File Not Open[2]	255: (control-C Interrupt)
19: Duplicate File Name[2]	
20: File Busy[2]	[1]DOS 3.3 only
21: File(s) Still Open[2]	[2]ProDOS only

DECIMAL		HEX
224-225	**X of Last HPLOT** (0-279)	$E0.E1
226	**Y of Last HPLOT** (0-191)	$E2
228	**HCOLOR Code** 0=**0**, 42=**1**, 85=**2**, 127=**3**, 128=**4**, 170=**5**, 213=**6**, 255=**7**	$E4
230	**Hi-Res Plotting Page** POKE 230,**32** selects Page 1. POKE 230,**96** selects Page 3. POKE 230,**64** selects Page 2.	$E6
231	**SCALE** Note: **SCALE=0** is equivalent to a SCALE of 256.	$E7
232-233	**Shape Table Start Address**	$E8.E9
234	**Hi-Res Collision-Check** Example: XDRAW a shape. If **PEEK(234)=0** then the shape started at a *non-black* hi-res point.	$EA
241	**SPEED** Note: **PEEK(241)** is 256 *minus* the current SPEED.	$F1
243	**FLASH Mask**	$F3
249	**ROT**	$F9

Display Switches

DECIMAL (with negative equivalent)		HEX
49232 (-16304)	**Graphics**	$C050
49233 (-16303)	**Text**	$C051
49234 (-16302)	**Full-Graphics**	$C052
49235 (-16301)	**Split-Screen**	$C053
49236 (-16300)	**Page One**	$C054
49237 (-16299)	**Page Two**	$C055
49238 (-16298)	**Lo-Res**	$C056
49239 (-16297)	**Hi-Res**	$C057

Note: Activate display switches by Poking each location.
Example: **POKE 49232,0** switches to Graphics display.

Keyboard, etc.

DECIMAL (with negative equivalent)		HEX
49152 (-16384)	**Read Keyboard**	$C000
49168 (-16368)	**Clear Keyboard** Example: 10 KEY=**PEEK(49152)**: IF KEY<128 THEN 10 20 **POKE 49168, 0** 30 PRINT "KEY: "; CHR$(KEY-128)	$C010
49200 (-16336)	**Click Speaker** Example: FOR A=1 TO 99: BUZZ=**PEEK(49200)**: NEXT	$C030
49249 (-16287)	**Button #0** Paddle-0 Button or Open (left) Apple key.*	$C061
49250 (-16286)	**Button #1** Paddle-1 Button or Closed (right) Apple key.*	$C062
49251 (-16285)	**Button #2** *Example: If **PEEK(49249+P)** is greater than 127, then Paddle Button #P is being pressed—or it's not connected.	$C063

DOS 3.3 Pokes

(assume DOS loaded in main memory)

POKE 40193, PEEK(40193)–N: CALL 42964
Moves DOS buffers down N★256 bytes.

POKE 44452,N+1: POKE 44605,N
Allows N file names before catalog pause.

POKE 44460,88: POKE 44461,252
Clears screen before catalog.

POKE 44505,234: POKE 44506,234
Exposes deleted file names in catalog.

POKE 44596, 234: POKE 44597, 234: POKE 44598, 234 Cancels catalog pause.

POKE 49107,234: POKE 49108,234: POKE 49109, 234 Prevents language card reload.

POKE 49384,0 Stops drive motor.

POKE 49385,0 Starts drive motor.

Notes

Apple's main memory consists of 65,536 *bytes*, numbered zero to 65535. Every byte has a *value* in the range 0-255.

- You may *Peek* (look at) the value in byte number-B with the command— **PRINT PEEK(B)**
- You can usually *Poke* a new value-V into byte-B with the command— **POKE B,V**

Values higher than 255 must be stored in two bytes:

- To look at the value in consecutive bytes B1-B2— **PRINT PEEK(B1)+PEEK(B2)*256**
- To Poke a new value V (0-65535) into bytes B1-B2— **POKE B1, V-INT(V/256)*256** and **POKE B2, INT(V/256)**

Note: Since almost any memory location can be Peeked or Poked, program listings can reveal thousands of Peeks and Pokes not listed on this chart. Pokes are often used to write machine-language routines that may be activated with the CALL command—the possibilities are *infinite*.

Let A=PEEK(64435) and B=PEEK(64448).
If A=6 and B=0 then Apple IIc.
If A=6 and (B>223 AND B<240) then Apple IIe.
If A<>6 then Apple II or II+.

Page-3 DOS Vectors

DECIMAL		HEX
976-978	**Re-enter-DOS Vector**	$3D0.3D2
1010-1012	**Reset Vector** Example: POKE 1012, 0 makes Reset boot. (POKE 1012,56 to restore normal Reset function.)	$3F2.3F4
1013-1015	**Ampersand Vector** Examples: POKE 1014, 165: POKE 1015, 214 makes "&" LIST. POKE 1014, 110: POKE 1015, 165 makes "&" CATALOG. POKE 1014, 18: POKE 1015, 217 makes "&" RUN.	$3F5.3F7
1016-1018	**Control-Y Vector**	$3F8.3FA

DOS 3.3 Locations

(All values assume DOS is loaded in main memory.)

DECIMAL		HEX
42350	**Catalog-Routine** Example: CALL 42350 catalogs a disk.	$A56E
40514	**Greeting Program Run-Flag** POKE 40514,52 and INIT a disk. When booted, DOS will attempt to *BRUN* the greeting program. POKE 40514,20 for *EXEC*.	$9E42
43140-43271	**Commands**	$A884.A907
43378-43582	**Error Messages**	$A972.AA3E
43616-43617	**Last Bload Length**	$AA60.AA61
43634-43635	**Last Bload Start**	$AA72.AA73
43624	**Drive-Number** Example: **POKE 43624, D** changes disk input/output to Drive D.	$AA68
43626	**Slot-Number** Example: **POKE 43626, S** changes disk input/output to Slot S.	$AA6A
43698	**Control-D Command Character**	$AAB2
44033	**Catalog Track Number**	$AC01
45991-45998	**File-Type Codes**	$B3A7.B3AE
45999-46010	**Disk Volume Heading**	$B3AF.B3BA
46017	**Disk Volume Number**	$B3C1

ProDOS™ Locations

DECIMAL		HEX
48944	**Slot/Drive Value** If **PEEK(48944)** is greater than 127 then Drive 2, otherwise Drive 1.	$BF30
47313-47422	**Commands**	$B8D1.B93E
48840-48841	**Last Bload Length**	$BEC8.BEC9
48825-48826	**Last Bload Start**	$BEB9.BEBA

Useful Calls

DECIMAL (add 65536 for positive equivalent)		HEX
CALL-25153	**Reconnect DOS 3.3**	$9DBF
CALL-3100	**Reveal hi-res page 1**	$F3E4
CALL-3086	**Clear hi-res screen to black**	$F3F2
CALL-3082	**Clear hi-res to last color Hplotted** Example: HGR2: HCOLOR=5: HPLOT 0,0: **CALL-3082**	$F3F6
CALL-2613	**Hi-res coordinates to Zero-Page** Example: The X and Y starting coordinates of the *next* shape table DRAW or XDRAW may be determined with a **CALL-2613**. Then X=PEEK(224)+PEEK(225)*256 and Y=PEEK(226).	$F5CB
CALL-1438	**Pseudo-Reset**	$FA62
CALL-1370	**Boot**	$FAA6
CALL-1321	**Display all registers**	$FAD7
CALL-1184	**Clear screen and print "Apple . . ."**	$FB60
CALL-1036	**Move cursor right**	$FBF4
CALL-1008	**Move cursor left**	$FC10
CALL-998	**Move cursor up**	$FC1A
CALL-958	**Clear text from cursor to bottom**	$FC42
CALL-922	**Move cursor down**	$FC66
CALL-868	**Clear text-line from cursor to right**	$FC9C
CALL-756	**Wait for any keypress**	$FD0C
CALL-678	**Wait for a Return keypress**	$FD5A
CALL-657	**Better Input; commas/colons o.k.**	$FD6F

```
10 PRINT "NAME (LAST, FIRST) :" ; : CALL -657
20 A$="" : FOR X=512 TO 767: IF PEEK(X)<>141
   THEN A$=A$+CHR$(PEEK(X)-128): NEXT X
```

CALL-468	**Memory move** A Basic memory move: OS & OE are the *Old*-location Start & End, and NS is the *New* Start. GOSUB 5000 to execute the move—	$FE2C

```
5000 N=OS: LOC=60: GOSUB 5020:
     N=OE: LOC=62: GOSUB 5020:
     N=NS: LOC=66: GOSUB 5020
5010 POKE 768, 160: POKE 769, 0: POKE 770, 76:
     POKE 771, 44: POKE 772, 254: CALL 768: RETURN
5020 POKE LOC, N-INT(N/256)*256:
     POKE LOC+1, INT(N/256): RETURN
```

CALL-415	**Disassembler** Note: Poke start address at locations 58-59 before Call.	$FE61
CALL-211	**Ring bell and print "ERR"**	$FF2D
CALL-198	**Ring bell**	$FF3A
CALL-151	**Enter monitor**	$FF69
CALL-144	**Scan input buffer** This example uses CALL -144 to execute a machine language routine from Basic (will not work in a subroutine):	$FF70

```
100 A$="300: A9 C1 20 ED FD 18 69 01 C9 DB D0 F6
     60 300G D823G"
110 FOR X=1 TO LEN(A$): POKE 511+X,
     ASC(MID$(A$,X,1))+128: NEXT
120 POKE 72, 0: CALL -144
```


6/84

in time with the buzzing. Who would think of such a thing? My nostalgia for this ephemera also owes to the tangibility and simplicity of these details. The Apple IIe was a very limited machine next to the Macintosh, which arrived only a few years later. With its text screen and minimal graphics, the Apple IIe offered a mechanical transparency that people these days obtain only from using Arduino circuit boards and working with firmware.

PCs at that time did not have multitasking, something that is so taken for granted today that imagining a computer without it seems absurd. Multitasking is the ability for a computer to run multiple programs at the same time. If you look at the Task Manager on Windows or the process table on Linux or OSX, you will see that your computer is running dozens if not hundreds of programs (or tasks, or processes) simultaneously, with the operating system's core "kernel" (the central controller of the entire operating system) allocating work to CPU cores in very intricate fashion. This scheduling is entirely opaque to users and to most programmers. But the Apple IIe, like most personal computers of the early to mid-eighties, did *one thing at a time*. If I asked Applesoft BASIC to **`PRINT "HELLO"`** and hit return,* the CPU would devote itself exclusively to printing **`HELLO`** on the screen until it was finished, at which point it would wait for further input from the user. Even the original Macintosh, with its graphical user interface, could not stop a program from running once it was executed (though it and MS-DOS both had mechanisms for tiny programs like a clock or a device driver—or a virus—to remain semi-present even if they weren't technically running). So at any time, only *one program could run,* and it did so on top of the operating system, which ran on top of the hardware CPU, the central processing unit.

Computers are best understood as a series of *abstraction layers,* one on top of the other. Each new top layer assembles the previous lay-

* The "Enter" key was exclusive to IBM computers. There is probably an interesting history about the concurrent evolution of "Enter" vs. "Return," to parallel the double command characters of "line feed" and "carriage return." Briefly, a carriage return moves the cursor position (where the next character is to be printed) back to the beginning of the line (the left, unless you're typing in Hebrew, Arabic, Thaana, N'Ko, Mende Kikakui, or a couple other scripts), while a line feed moves the cursor to a new line. On a typewriter, line feed meant "down" and carriage return meant "go to the beginning of the line," but once computer screens replaced printing typewriters, the two coalesced to carry the same meaning: go down a line and back to the left (or right).

er's pieces into more complex, high-level structures. On the Apple, a BASIC program can be the top level, which executes on top of the DOS operating system, which executes on the hardware. The bottommost level is the hardware: the CPU. The CPU consists of more than a billion transistors arranged so that they can physically execute an "assembly" language (or "machine language") that is native to that CPU. Assembly is the deepest layer of code, where one can directly give the CPU instructions. And what one can tell it to do is often pretty limited: store this number here, retrieve this number from there, add or subtract these two numbers, and branch to different bits of code depending on some condition or other. In different contexts, these operations can take on different meanings, such as printing text onto a screen or sending something across a network, but the overall level of *structure* is very primitive. Assembly can be tedious and even painful to program in—but because it is the language of the CPU, it is fast.

It's often said that an algorithm is a recipe, but let's extend that analogy to computing abstraction layers, to illustrate how assembly connects to the hardware beneath it and the high-level language (BASIC, Logo, or something else) above it. Imagine that we're in a restaurant. A BASIC program is a diner. Diners read the menu and know just enough about the dishes to decide what they want to eat. They don't have to worry about how much salt to put in the soup, how long dishes need to cook, or how to lay out the food on the plate. They have only *high-level* control over the end result, their meal. Without knowledge of what goes on in the kitchen, diners take into account their taste preferences, allergies, recommendations, and such, and order a meal off the menu. The chef, who knows how to translate the dishes on the menu into actual recipes, is the *compiler* or *interpreter*, who translates high-level instructions (the diner's order) into far more specific low-level instructions (the exact ingredients list and instructions for cooking). The kitchen cooks are the actual computing hardware, who have the expertise to perform a variety of precise cooking skills reliably without error. They make the dishes based on the exact instructions given to them by the chef. The diners remain ignorant of the details.

High-level BASIC programs were translated (or interpreted) into a language called 6502 assembly for Apple IIe CPUs. The clock speed of a processor, given in cycles per second, or hertz, dictates just how fast a CPU chip could execute individual assembly instructions.* Vastly more daunting than BASIC, I didn't dare touch 6502 assembly as a kid. Assembly language grants access to the physical memory of the computer and allows one to specify numerical operation codes (opcodes) that are actually understood by the hardware in the CPU. In assembly, there is almost no distance between the programmer and the hardware.

Here's some assembly for a "Hello world!" program (one that just displays "Hello world!" and exits) in Apple II 6502 assembly:

```
COUT  gequ   $FDED           ;The Apple II character output func.

      keep   HelloWorld

main  start
      ldx    #0              ;Offset to the first character
loop  lda    msg,x           ;Get the next character
      cmp    #0              ;End of the string?
      beq    done            ;->Yes!
      jsr    COUT            ;Print it out
      inx                    ;Move on to the next character
      jmp    loop            ;And continue printing
done  rts                    ;All finished!
msg   dc     c'Hello world.'
      dc     h'0D'
      dc     h'00'
      end
```

* The 6502 was a 1 megahertz processor: it operated at one million clock cycles per second. It allowed for the execution of a bit less than half a million assembly instructions during that time. Sometime after the year 2000, increasingly sophisticated design, as well as the advent of multiprocessor machines and dedicated graphics processing units, ceased to make clock cycles a meaningful indicator of performance, so terms like a 3 gigahertz processor mean less today because their speed at executing instructions (as well as the content of those instructions) can vary wildly. In 2003, the classic 3 gigahertz Pentium 4 processor could execute about ten billion instructions per second. By 2012, a 3.2 gigahertz Intel Ivy Bridge chip with four CPUs (or cores) could theoretically peak at executing 130 billion instructions per second, over 30 billion per core, though avoiding the many other possible bottlenecks to reach that speed was not trivial. It's probably still fair to say that that 2012 Ivy Bridge chip is about 100,000 times as powerful as the 6502 inside my old Apple IIe. What has been utterly lost is the linear nature of the 6502, which executed one instruction at a time in strict sequence. Processor development has gradually relaxed the idea that a computer is a calculator performing strictly ordered operations at increasing speeds.

And here it is in C:

```
int main() {
     printf("Hello world!\n");
     return 0;
}
```

And here it is in Applesoft BASIC:

```
10 PRINT "HELLO WORLD!"
```

In the eighties, many programmers coded directly in assembly. Programs were simpler and performance was critical. But as computers got larger and more complex, it became unfeasible to code in assembly.* Programmers need to learn a different assembly language for different processors (as with the Apple II's 6502, the Macintosh's 68000, and the PC's 8086), which is horrendously inefficient. More efficient was to use a CPU-independent higher-level language. All the languages we hear about today, from C++ to Java to Ruby to Python, are higher-level languages. A compiler takes the code written in these languages and translates it into the assembly code for a particular processor.

Until I learned assembly in college, and how language compiler programs translated higher-level programming languages into assembly, computers remained partly opaque to me. That gap in my knowledge bothered me, because even though I had far more direct control over those lower layers, I couldn't understand them. When I took a compilers class in college, the infrastructure of the computer opened up to me. There was no longer a miracle in between my code and its execution. I could see the whole picture, finally, and it was beautiful.

* John von Neumann, one of the greatest geniuses of the twentieth century and the inventor of the standard architecture that forms the basis of nearly all modern computers, was so fluent with assembly that he saw no need for higher-level languages whatsoever. In 1954, thinking the whole idea a waste of time, he said, "Why would you want more than machine language?" He did not foresee, as almost no one did, a time when programs would grow so large even in these higher-level languages that they would defy full comprehension by a single person.

The Split

> I renounce any systematic approach and the demand for exact proof. I will only say what I think, and make clear why I think it. I comfort myself with the thought that even significant works of science were born of similar distress.
>
> I want to develop an image of the world, the real background, in order to be able to unfold my unreality before it.
>
> —ROBERT MUSIL

When I was a teenager, programming lost its allure. The "real world," such as it was, had drawn my attention away from what now looked to be the sterile, hermetic world of computers. It was the late eighties. The web did not exist in any accessible form, nor were computers part of most people's daily lives. I was part of the very last generation to grow up in such a world. People only a few years younger than me would have the nascent public internet and the web to dig into and explore. I had online bulletin board systems (BBSs) and such, but they were strictly cordoned off from my everyday existence, the exclusive preserve of hobbyists, eccentrics, and freaks. And I was miserable in my small suburban enclave. For many programmers, computers held the answer to such misery. They continue to provide the mesmeric escape from the dreary everyday routines of teenage and adult years. I don't have a clear explanation as to why computers failed to offer me solace as they did for many others. Something kept me from locking in completely to the brain-screen bond that kept many teen programmers up all night coding games or hacking copy protection. Literature became my refuge instead.

My parents had raised me on science fiction, the standard literary junk food of computer geeks, but I felt increasingly drawn to explorations of human emotion and existential crisis. At a point of typical thirteen-year-old despair, I devoured the complete works of Kurt Vonnegut over the course of two weeks. They touched me. Vonnegut led me to explore increasingly "deep" fiction.*

* The teenage definition of "deep" being "full of angst and weltschmerz." The three big authors then for budding, alienated youth were J. D. Salinger, Albert Camus, and Sylvia Plath. I dutifully read them all.

My high school physics teacher introduced my class to James Joyce, whom he considered the greatest author of the twentieth century. He told us that *Ulysses* was dauntingly complex and that *Finnegans Wake* was simply incomprehensible. The difficulty and obscurity of *Ulysses* intrigued me as a teenager much as that Logo tree program had as a child. How could a book of fiction be "difficult"? Did it too hold a kind of programmatic complexity to it?

I had stumbled on the writers of the Oulipo, the French-dominated group specializing in experimental works of "potential literature," after Martin Gardner, amateur mathematical enthusiast, had published several articles on the group in his column in *Scientific American.** Their most famous members—Raymond Queneau, Georges Perec, Italo Calvino, and Harry Mathews—specialized in the innovation of literature through the use of formal constraints. One of the most infamous was Perec's novel *La disparition* (*A Void* in Gilbert Adair's English translation), which contains not a single *e* in its three hundred pages. That kind of constraint—leaving out a letter or set of letters—is called a lipogram. The poet Jean Lescure's "S+7" method replaces every noun in a text with the seventh noun following it in the dictionary. "Lend me your ears" becomes "Lend me your easels." Raymond Queneau's *One Hundred Thousand Billion Poems* is a set of ten sonnets with the exact same rhyme scheme and rhyme sounds. By mixing and matching lines, there are ten possible first lines, ten possible second lines, and so on, resulting in 10^{14} possible sonnets.

Georges Perec's mighty *Life: A User's Manual* describes the inhabitants of the ninety-nine rooms of a 10x10 apartment building (one corner is missing), where each successive room is a chess knight's move away from the current room, and each room is visited only once.† The intricate structuring, I later discovered, had close ties to computer science. Perec was fascinated by a mathematical technique called the Graeco-Latin square, a device that pairs up two sets of elements so that

* Gardner deserves a great deal of credit for being one of the great popularizers and connectors of mathematically inflected art and recreations in the twentieth century. He was a one-man archive of nerd high culture, and possibly helped create more young mathematicians and computer scientists than any other single factor.

† Perec's path through the apartment building is called a Knight's Tour. It is commonly assigned in introductory computer science classes: program a computer to find such a Knight's Tour, if one exists, given a board of some size. Perec had to find his path by hand.

each pair occurs only once. The pairs are distributed in a square so that each element also occurs only once in every row and column. Perec used (and abused) Graeco-Latin squares to structure his works: the original plan for *Life: A User's Manual* was to utilize over twenty squares to determine what objects, characters, times, furniture, clothing, and music to place into each chapter. Finding these squares was a devilish task: in the eighteenth century, Euler thought no 10x10 Graeco-Latin squares existed, and one was only found in 1959, with the assistance of a computer.

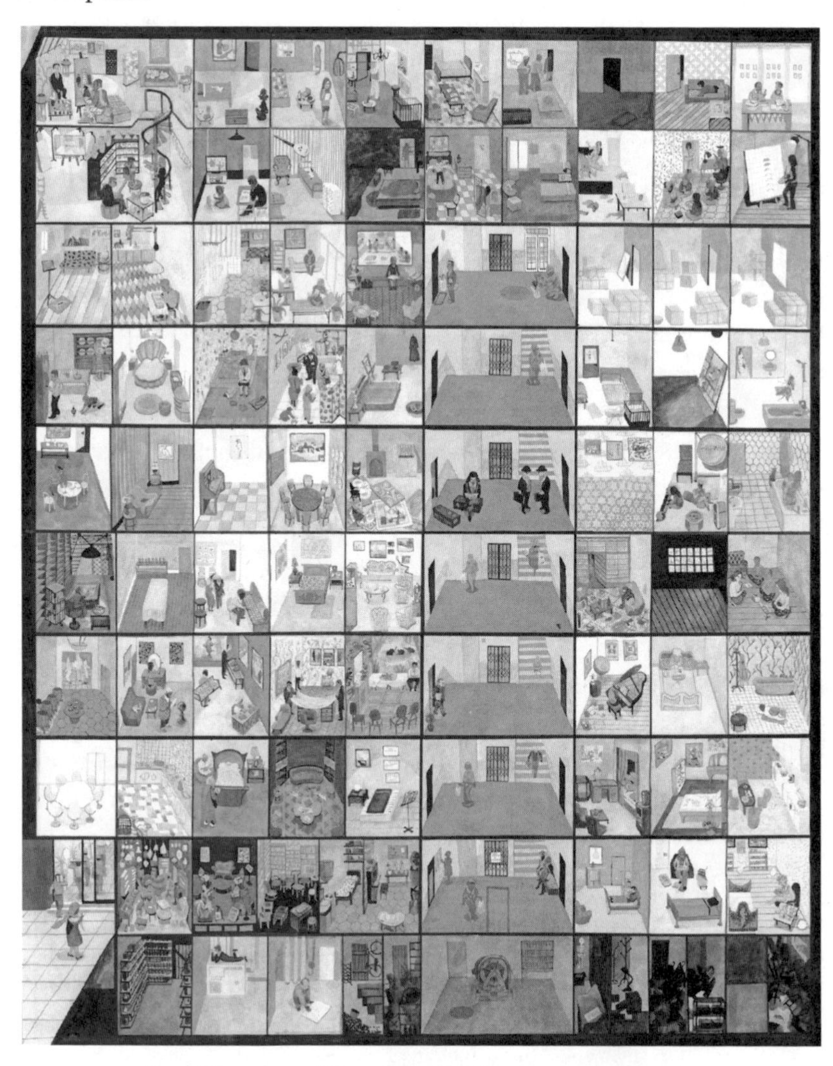

Brecht Evens's rendition of the 10x10 apartment building in Georges Perec's *Life: A User's Manual.*

Aα	Bδ	Cβ	Dε	Eγ
Bβ	Cε	Dγ	Eα	Aδ
Cγ	Dα	Eδ	Aβ	Bε
Dδ	Eβ	Aε	Bγ	Cα
Eε	Aγ	Bα	Cδ	Dβ

A 5x5 Graeco-Latin square. Each pair occurs only once, and each symbol occurs only once in each row and column.

Perec, unable to construct squares of sufficient sizes himself, wrote to Indra Chakravarti, one of the authors of the 1960 paper "On Methods of Constructing Sets of Mutually Orthogonal Latin Squares Using a Computer," who provided Perec with two 12x12 squares. One of Chakravarti's coauthors was computer scientist Donald Knuth, who would go on to write the monumental bible of algorithmic programming, *The Art of Computer Programming*. Knuth calls *Life: A User's Manual* "perhaps the greatest 20th century novel." Perec could not have written some of his works without the help of a computer. Knuth found a majestic novel in the combination of algorithmic constraints and Perec's protean creativity.

The *mathematical* appeal of constraints like the Knight's Tour and the Graeco-Latin square lies, I believe, in how precisely specified and quantified they are. An author can write "a science-fiction story" or a "third-person narrative" or a "prison narrative," but these are only the most general terms for vaguely defined genres. There are no hard-and-fast rules for what qualifies as a particular type of story. Oulipian constraints, like the lipogram or S+7, are fixed and firm, and therefore *algorithmic*—most of the time. Raymond Queneau's *Exercises in Style* consists of ninety-nine different retellings of a banal anecdote about two men encountering each other on the bus and then on the street. Some versions are purely formal, permutations of letters and words according to precise algorithms, but many are stylistic variations in slang, music, descriptive language, dialogue, and other modes of expression. They are not wholly formal maneuvers; the book is formally organized to spur impulsive, creative expression under posited constraints. Comic artist Matt Madden later reproduced the conceit with different visual styles in *99X*. I found Queneau's looser stylistic rules more compelling than the purely formal permutational algorithms. My life did not resemble the fractal pattern of that Logo tree, for better or for worse. It could not be generated or replicated by a simple algorithm. The Oulipo captivated me because the movement's members incited creativity, not because they eliminated it.

When I encountered the Oulipians, I felt the first tug between what I'll call the *analytic* impulse and the *heuristic* impulse. The analytic impulse finds its fulfillment in the sheer beauty of formal maneuvers. A perfect algorithm, a beautiful mathematical proof, a meticulously organized table of contents: structure, coherency, and elegance are the motivations that drive analytically inclined people to accomplishment. Yet these systems were too detached from the messiness of "real" "life" to absorb me fully. In contrast, there is the *heuristic* approach, by which we make provisional and approximate estimates in order to deal with intractable complexity and ambiguity, drawing from our gut instincts, our past experience, our subconscious, and other mental processes that we ourselves may not fully understand. The modern sense of the term "heuristic" was coined by mathematician George Pólya, who described heuristics as possessing both logical and psychological aspects, inextricably tied together. Pólya deemed them integral to mathematical thinking, but their use goes far beyond that. The nuances of human language and emotion, the difficulties of justice and politics, and the profundity of aesthetic experience are phenomena that have resisted being neatly quantified into the sort of analytic boxes that mathematics and logic provide. So rather than perfect an analytic representation, we accept inaccuracy and error.

These kinds of opposing tendencies have been noticed by many. Neither side is specific to science or the humanities. Both the analytic and the heuristic exist within any domain of study, whether literature, logic, or sport. In 1905, mathematician, astronomer, and writer Henri Poincaré distinguished two types of mathematicians:

> The one sort are above all preoccupied with logic; to read their works, one is tempted to believe they have advanced only step by step, after the manner of a Vauban who pushes on his trenches against the place besieged, leaving nothing to chance. The other sort are guided by intuition and at the first stroke make quick but sometimes precarious conquests, like bold cavalrymen of the advance guard. . . . Logic, which alone can give certainty, is the instrument of demonstration; intuition is the instrument of invention.

Intuition, Poincaré says, is both necessary and fallible. I call it "heuristic" because it is the often-unconscious art of selecting which facts are relevant, which phenomena are linked, and which shortcuts to take. Seventy years later, mathematician Mark Kac spoke of two species of genius, scientific and otherwise: the "ordinary" and the "magician." The difference, Kac says, is that even after we understand what a magician like Richard Feynman has done, we still have no idea how they got there.

Heuristics may seem like inferior mental shortcuts compared to the exactness of an algorithm, but in the 1950s, Nobel economic laureate and polymath Herbert Simon promoted heuristics as a necessary tool for coping with a world too complex to understand analytically. For Simon, heuristics were a necessary mechanism for dealing with the limitations we face in solving any problem: limitations of time, of knowledge, of brainpower. Psychologist Gerd Gigerenzer goes further, emphasizing that a failed complex analysis often generates *worse* results than a simple, heuristic decision. We bring our biases to problems not because we are flawed but because we would be utterly lost without them: "Without bias, a mind could not function well in our uncertain world."

Heuristics are indeed necessary for human functioning, but we must be cautious in how we apply them and their biases. As we'll see later, when heuristics are carelessly translated to computers, trouble follows.

Bits and pieces inevitably slip through the cracks of heuristics. Those lost fragments, too complex to be captured by formal analysis or heuristic shorthand, fascinated me as much as the formal systems. When Oulipian writers wove formal abstractions into human joy and grief, as

Jacques Roubaud did in *The Great Fire of London* and Perec did in *W, or The Memory of Childhood,* they brought out the gaps between those abstractions and the irreducible complexity of reality. The formal and the analytic, in their hands, became a heuristic tool in itself. Oulipian techniques offered me unorthodox tools for connecting with the world of human will and emotion.

The Oulipians played great games as well, but sometimes the game seemed to take precedence over the human significance of the story. If one is going to generate stories out of an arrangement of tarot cards, as Italo Calvino did in *The Castle of Crossed Destinies,* will the results pierce the human heart? At times yes, at times no, and so I found myself increasingly drawn to the less constrained writing of Virginia Woolf and Herman Melville.

But it was at the station of James Joyce where I moored my boat. I read *Ulysses* with a brilliant and sympathetic teacher. The book was damnably hard,* with incomprehensible passages of Irish dialect, Catholic theology, and gutter obscenity. Joyce's ideas ranged from puerile to abstruse to profound, freely mixed together with no easily grasped logic. Joyce's book was meticulously written, yet his plan remained opaque to me. That challenge kindled a similar curiosity in me as Logo had done years prior. But while computational concepts, however difficult, resolved themselves clearly, Joyce's *Ulysses* opened itself up to a myriad of interpretations. The characters of *Ulysses*—Stephen Dedalus, Molly Bloom, and Leopold Bloom—possessed lives that were laden with tragedy, loss, and pain. Stephen's loss of his mother, the Blooms' loss of their son, and the wayward wandering of their daughter—to me they were matters of the highest importance.

Yet for all of *Ulysses*'s rich messiness, it had been rigorously structured, even overstructured, by Joyce. Joyce distributed several schemas purporting to lay out the plan of the book, describing chapter-by-chapter parallels with episodes of the *Odyssey,* as well as the symbols and organs of the body that dominated each chapter. Yet Joyce's own words make it clear that the schema is not the be-all and end-all of

* I was not, in fact, a snob or a would-be snob. I was simultaneously consuming gargantuan helpings of trash in the form of comic books, movies, television, and middling fiction. Most of this trash hasn't stayed with me (but neither has most of the highbrow stuff), which I think is the real marker between great works and mediocre ones.

the book. It was only one way (or eight ways) of seeing the novel, and Joyce had worked himself to exhaustion to ensure that no one interpretation or analysis could be final. The overlaid structures contradicted one another. A character could be a hero or a villain, or a success or a failure, depending on what prismatic structure the reader applied. No single one was correct. This was Joyce's way of drawing out what was lost in those gaps, by providing not one but *many* conflicting heuristics for understanding the book. Joyce's goal, as I came to see it, was not just to leave in ambiguity but to pile on contradiction upon contradiction. To enrich rather than reduce.

The formal patterns of *Ulysses* were fascinating to me, but not in and of themselves. Rather, they disguised and then revealed clues to the deepest puzzles of existence, those half-shown to us in dim light as the knotted tendrils of human feeling. Computers could not compare.

The Join

> It is only a frivolous love that cannot survive intellectual definition; great love prospers with understanding.
>
> —LEO SPITZER

I met my wife Nina when I was eighteen. We wouldn't get married for ten years, because who could possibly trust their eighteen-year-old self to be competent at choosing a partner? Our meeting had been a freak accident. I was visiting friends at Harvard, who were hosting an end-of-Passover pizza dinner at Pizzeria Uno. Seven or eight of us squeezed around a small table, and the person next to me was Nina. Nina wanted to be a poet. I wanted to be a novelist. We were both programmers. Nina and I each decided that the other was more interesting than anyone else at the table. I did not see her again for six months.

Over the summer, we exchanged emails daily while I did data entry and programming for a factory that made self-locking fasteners. We exchanged adolescent angst and book and movie recommendations. She liked the Cocteau Twins. I liked the Gang of Four. We made cassette mix tapes for each other. We eventually met up again and got

	TITLE	SCENE	HOUR	ORGAN
1.	Telemachus	The Tower	8 a.m.	
2.	Nestor	The School	10 a.m.	
3.	Proteus	The Strand	11 a.m.	
4.	Calypso	The House	8 a.m.	Kidney
5.	Lotus-eaters	The Bath	10 a.m.	Genitals
6.	Hades	The Graveyard	11 a.m.	Heart
7.	Aeolus	The Newspaper	12 noon	Lungs
8.	Lestrygonians	The Lunch	1 p.m.	Esophagus
9.	Scylla and Charybdis	The Library	2 p.m.	Brain
10.	Wandering Rocks	The Streets Room	3 p.m.	Blood
11.	Sirens	The Concert	4 p.m.	Ear
12.	Cyclops	The Tavern	5 p.m.	Muscle
13.	Nausicaa	The Rocks	9 p.m.	Eye, Nose
14.	Oxen of the Sun	The Hospital	10 p.m.	Womb
15.	Circe	The Brothel	12 midnight	Locomotor Apparatus
16.	Eumaeus	The Shelter	1 a.m.	Nerves
17.	Ithaca	The House	2 a.m.	Skeleton
18.	Penelope	The Bed		

One version of the patterns of *Ulysses,* as set down by Joyce.

	ART	COLOUR	SYMBOL	TECHNIC
1.	Theology	White, gold	Heir	Narrative (young)
2.	History	Brown	Horse	Catechism (personal)
3.	Philology	Green	Tide	Monologue (male)
4.	Economics	Orange	Nymph	Narrative (mature)
5.	Botany, Chemistry		Eucharist	Narcissism
6.	Religion	White, black	Caretaker	Incubism
7.	Rhetoric	Red	Editor	Enthymemic
8.	Architecture		Constables	Peristaltic
9.	Literature		Stratford, London	Dialectic
10.	Mechanics		Citizens	Labyrinth
11.	Music		Barmaids	*Fuga per canonem*
12.	Politics		Fenian	Gigantism
13.	Painting	Grey, blue	Virgin	Tumescence, Detumescence
14.	Medicine	White	Mothers	Embryonic
15.	Magic		Whore	Hallucination
16.	Navigation		Sailors	Narrative (old)
17.	Science		Comets	Catechism (impersonal)
18.	Flesh		Earth	Monologue (female)

together. Both of us had grown up as science and math geeks, yet both of us were enamored of literature and emotional quandries. She gave me James Agee and Walker Evans's *Let Us Now Praise Famous Men*. I gave her Jorge Luis Borges's *Ficciones*. We looked for the supposed plate memorializing William Faulkner's Quentin Compson on the Charles River Bridge in Cambridge. We never found it, though we did nearly freeze.* A friend told me she knew Nina was special because she had convinced me to wear rainbow shoelaces. She had a far better ear for language and music than me, and she made me see beauty where I had only been looking for rigor and strain.

If there was an algorithm for our feelings, it had to be reverse engineered. Our emotions and reactions often seize us with so much force that they wipe away any possibility of detached cognition. Love is irrational, it is passionate, it is madness. When I met Nina, I was scared. My previous relationship had ended badly and left me shell-shocked. It took months, if not longer, for me to separate Nina from my past.

There is a standard progression to many relationships: There is the initial crush, during which we are quick to overlook the flaws of the other person. When the rush of hormones and infatuation fades, we see what we previously couldn't. The other person reveals their flaws, their peccadilloes, and their small failures. Then begins the real work of negotiating. We remember the feeling of the initial crush. Maybe we wonder how things changed and why it is that the other person can no longer provoke that same level of positive joy. If we're cynical, we see that our minds become high on chemicals and gratifying delusions—tiny and harmless perversions of the true picture of the other person.

Dating and relationships give us a personal, private history of memories and associated feelings. From that, we construct reductive judgments. We sum up the incomplete data we have about our partners into heuristic measures: how loving they are, how smart, how dutiful, how talented, how promising. We recalculate constantly. Our willingness to continue the relationship depends on how positive those judgments continue to be.

But do we make these judgments with any consistency or rationality? Can we ever be sure we are weighing the evidence carefully enough?

* There are now several online guides to finding the small and unobtrusive plaque, none of which existed then. It reads "Quentin Compson III. June 2, 1910. Drowned in the fading of honeysuckle."

My wife and I waited ten years to get married.* When I read the storybook romances of the *New York Times* wedding section, I think: the plural of anecdote is not data.† Just as bugs lurk in the most beautiful of code, many loving marriages stand a good chance of ending in divorce. Inauspicious data kept us from getting engaged, and it would have surprised me then to know I was entering a relationship that has now spanned over half my life.

For anyone who uses the phrase "Till death do us part," marriage is predicated on the certain belief that it will be permanent. Mistrustful of emotions, especially my own, I was determined to make my decision based on evidence, not feeling. I did not know then what love meant, and whether it even meant the same thing to different people. Love as I conceived of it was less a matter of *feeling* love than *defining* it. Love was what made a successful relationship possible.

After college, I worked as a software engineer, first at Microsoft and then at Google. At night, I busied myself reading literature and writing. I also attended to the hard work of a relationship. I worked because my feelings and intuitions were insufficient. I wanted the hard verification of experience. I said to myself: My partner and I are both software engineers. We code and build programs. One of those programs is our marriage. The code for our marriage is made up of our words, our actions, and our thoughts. Our job is to prevent bugs in our code from crashing the marriage.

We could not avoid the bugs. Every programmer makes them. Instead, we tried to anticipate and mitigate the damage. Robust code can survive unexpected inputs and circumstances. But often bugs will cause a crash, heralded by a message known to anyone who's ever programmed in Linux or Unix: **`Segmentation fault (core dumped)`**. The same software engineering maxims that saved my code have also saved our marriage. Here are seven.

* I read Proust's *In Search of Lost Time* almost immediately after getting engaged. It was fortunate I didn't read it earlier, as one of the main themes is that there is no continuity in our feelings toward another. We can snap in and out of love at a moment's notice. Had I read it before the engagement, it might have spooked me off marriage for another ten years.

† Political scientist Raymond Wolfinger coined the inverse of this phrase, "The plural of anecdote *is* data," around 1970. The shift, I believe, brings out the underlying meaning better: while standardized anecdotes in sufficient number can constitute good data, it's rare for a random assortment of anecdotes to rise to that level.

1. AVOID VAPORWARE.

Many products have crashed on the shoals of ambitious announcements that proved impossible for programmers to fulfill. Delays in shipping a new product could last years—occasionally decades—and the end result, if it is in fact realized, is inevitably a disappointment. The pressure to deliver "vaporware" puts immense strain on programmers, since they fear they can't live up to what's been promised—a legitimate worry. What's the ultimate in relationship vaporware? "Till death do us part." Nina and I removed that line from our marriage vows and replaced it with quotes from Robert Musil ("Love is the most talkative of all feelings") and James Wright ("Somewhere in me there is a crystal that I cannot find alone"). Another way of putting this maxim is: underpromise and overdeliver. I often tell my wife that I would far prefer to exceed her expectations than disappoint her. I would rather have her expect too little of me than too much.

2. BETA-TEST.

We were young and callow when we got together. So were our feelings. The code I wrote when I was twenty-one was good, but it was also naive. Our relationship survived not because of who we were then, but because of what we learned in the years following. We understood this sufficiently to mistrust the feeble and inept judgments of our younger selves. We waited a decade to get married, to accumulate enough evidence that the code was now robust enough to keep the product running smoothly. But even foundations do not last forever. Regular maintenance and upgrades are crucial, lest a once-healthy system decline into a creaky machine.

3. BUGS NEVER DISAPPEAR.

They only hibernate. Our worst fights, and in particular our most trivial fights, always came after nine p.m., though it took us ages to figure that out. One of us—and if I'm being honest, it was me far more often than it was Nina—would get bent out of shape over an unkind word, a logistical screwup, or some other domestic misdemeanor. Some friend was annoying, some bill needed paying, or one or the other of us had failed in one of those very specific ways that only has meaning within the long-established habits of a relationship. These fights, which always began with some tiny offense and inflated into competing indictments of how the original dispute stood for some bigger problem, were dumb. Raw emotion kept the momentum going past any sense of rationality and perspective. And they always started after nine p.m., running on the fumes of fatigue and confusion. After each one was settled, we bemoaned the waste of our voices and our nerves. Nina figured out one solution, which was just to walk out. Then I figured out the other, which was not to discuss anything too heavy, or even make a pointed criticism, after nine p.m. Over a decade down the line, that little discovery has probably saved us hours of wear and tear on our cortisol levels and amygdalae.

Bugs can seem evanescent. I saw server crashes that appear out of nowhere and, just as mysteriously, seem to disappear. Bugs *never* disappear. If you haven't fixed it, it's a dead certainty the enigmatic bug will return. Bad fights may abruptly dissipate for calmer times, but underlying issues fester, only to explode later if they aren't excavated.

4. FOLLOW THE 90/10 RULE.

This is an old law of software optimization: "A program spends 90 percent of its time in 10 percent of its code. Humans are very bad at guessing where that 10 percent lies." Early on, I thought that the strength of our relationship rested on our shared interests (literature, software) and our physical attraction to each other. Looking back, those turned out to contribute a lot less to relationship stability than agreeing on finances, sharing the same lifestyle, and appreciating sarcastic humor. Daily expressions of appreciation matter precisely because they are daily. Remembering not to wear my shoes indoors has done more to maintain my marriage than did our honeymoon. Nina's smartwatch, which vibrates against her wrist even when her phone is buried in her bag, solved a prosaic but chronic source of stress. Practical matters were crucial, as was breathing space. Each of us needed a room of our own where we could be alone.

A person has only a limited number of resources to devote to a task, even to a relationship. We tried not to spend them arguing over television shows or each other's eccentric habits. This is unnecessary optimization, and as comp-sci demigod Donald Knuth once said, "Premature optimization is the root of all evil."

5. DO THE GRUNT WORK.

The 9:1 proportion recurs in computer science. Computer scientist Mary Shaw wrote, "Less than 10% of the code has to do with the ostensible purpose of the system; the rest deals with input-output, data validation, data structure maintenance, and other housekeeping." That is to say, what *seems* like the core infrastructure of a relationship—love, sex, children—is made possible only through the support of auxiliary but equally essential pieces. Most of a relationship's code will not be about love, but about supporting the *possibility* of love, and that has to do with working out the piddly things such as scheduling and finances and chores. Without the necessary support infrastructure, a relationship will crash like early Windows. Beware software engineers *and* lovers who want to be there only for the sexy stuff.

6. HYSTERESIS ENABLES HOMEOSTASIS.

Hysteresis is the dependency of a system not just on its current state but also on its past states. Hysteresis is incorporated into an algorithm to prevent changes from having too many drastic effects too quickly. For example, a thermostat set to 68 degrees might not automatically heat at 67 degrees and cool at 69 degrees; it tracks recent activity to prevent itself from oscillating wildly.

In a marriage, too, it can be disastrous to overreact to sudden changes. The human mind has a great knack for rewriting the past based on the caprices of present emotions. In my worst moods, calm nights at home can abruptly look like the wastage of many years—every book read or movie watched just another distraction from the empty spectacle of two vacant souls enduring a pointlessly fortunate life. In our worst weeks—I count perhaps half a dozen of them—neither of us could see a way to continuing our relationship, and whatever was positive about our previous years together vanished from our memory's landscape. Those moods and memory lapses are also bugs. "With the perturbations of memory are linked the intermittencies of the heart," wrote the great computer scientist Marcel Proust. And with the perturbations of code are linked the intermittencies of software.

7. FAULT TOLERANCE.

Hidden bugs can come out of the woodwork years or decades after the code was written. (The Y2K bug, despite its lack of apocalyptic consequences, didn't show up for *decades* in systems that were designed in the 1970s and 1980s.) I once found a bug that would crash an entire chat server, kicking thousands of users offline, if a single user happened to send *exactly* sixty messages in one minute. Uncommon, unexpected inputs can produce catastrophic results, and the causes are rarely clear.

Good programmers realize that their knowledge of their programs is incomplete. They ensure that their programs are *fault tolerant* to the best of their abilities. After twenty years together, I know that my understanding of my wife is only an approximation of a complex individual. I have a chance of fully understanding a small enough segment of code to speak for its perfection; I could never say the same of a relationship. The formation of a relationship welds together disparate and incompatible pieces, each understood only approximately, into an indistinct structure requiring constant upkeep. We prepare not only for expected failures, but for unexpected ones. The relationship must be fault tolerant.

In 2000, back in the Triassic age of cellphone technology, my wife and I went on a disastrous kayaking trip organized by a friend indifferent to safety concerns. (I later found out his "expeditions" were called "death marches" by his other friends.) After paddling through the choppy and polluted waters of Puget Sound, just outside Seattle, we ended up stranded in darkness on a tiny spit of land with the tide encroaching. My kayak apron had not covered me correctly, and my clothes had gotten soaked. I shivered with incipient hypothermia while our friends bickered over which island on the map we were actually standing on. My wife quietly turned on the mobile phone that she'd presciently removed from its cobwebbed home in our car's glove compartment. She whispered to me, "We have cellphone reception. We can call the Coast Guard." I wouldn't have expected her to be able to reassure me in such a dire situation, but something in our shared history made us more robust, and more able to endure such an unexpected and bizarre experience, than either of us had anticipated.

Such moments mean more than opportunities for gratitude; they reassure me that the two of us have built a foundation that could endure many stresses both known and unknown. All we have, finally, is the wonder of the program itself—everyday life and love—and the unceasing task of having to debug and maintain our own code.

The Work

Quisque suos patimur Manis [Each of us suffers his own peculiar ghost]

—VIRGIL

Discovering compilers and the greater depths of algorithms, I renewed my enthusiasm for computer science in college. But that was not what caused me to leave the humanities. Rather, I became disenchanted with literature as a discipline of study as computers took on a brighter shine.

Upon entering college, my relationship with the humanities—founded in the rewarding study of James Joyce, Virginia Woolf, Ralph Ellison, and Herman Melville during high school—went sour quickly for me. My teenage fantasy of academia as refuge—a fantasy considerably more fixed than any ideal I had of a human relationship—was badly squelched. I had a handful of inspiring teachers who did open up the world of the humanities in new and profound ways.* But they only made more apparent that much humanities work, while serious in its way, was not serious *about literature*. The academic world was professionalized, competitive, and sterile. The evisceration of that scholarly dream was best put by John Williams in one of the greatest of all "academic" novels, *Stoner*.

> Stoner looked across the room, out of the window, trying to remember. "The three of us were together, and Dave said—something about the University being an asylum, a refuge from the world, for the dispossessed, the crippled. But he didn't mean [department chair] Walker. Dave would have thought of Walker as—as the world. And we can't let him in. For if we do, we become like the world, just as unreal, just as . . . The only hope we have is to keep him out."

Dreams stick as stubbornly as barnacles to our memories and feelings, so that they can only be removed once our minds have been

* In particular, literary scholar William Flesch, philosopher David Pears, the writer John Crowley, and the classicist Heinrich von Staden. After college, I found two more in the Joyce scholar Edmund Epstein and philosopher Galen Strawson.

rubbed raw. Such pain is the price, still a worthwhile one, for thinking to remain a serious task.

Computer science, while it became my undergraduate major and then the focus of my career, was *not* serious. I say that without irony: it was a free realm of play where symbols devoid of meaning were electronically juggled at superhuman speeds. Computer science was stable, secure, seemingly immune to trends, and most certainly on the rise. During college, I did an NSF fellowship in compiler research one summer, then interned at Microsoft the next. The promise of financial viability, work with real-world impact, and simple recognition and collaboration was far more appealing than the rat race of academia in which the slices of cheese were growing slimmer and slimmer. Where the humanities had been poisoned by becoming an occupation rather than a passion, software engineering was perfectly suited to disinterested professionalism and playful enthusiasm.

Ironically, the workaday world of software engineering, where I was forced to confront the perplexities of treating the world as data, gave me more occasion for philosophical reflection than the humanities did. It turned out I was ahead of the curve. There wasn't much data to be had in the late 1990s, but as I worked on internet services at Microsoft and then the far larger data silos of Google, I was confronted each day with the sheer strangeness of how software companies and software engineers were looking at the world: not quite as machines, but certainly not as most humans do. In this new realm, human language and human life were subjugated to the order of the binary.

2

CHAT WARS

Interop

It's easier to ask forgiveness than it is to get permission.

—GRACE HOPPER

AT TWENTY-TWO, when I was just out of college and still a green software engineer, I fought America Online, and AOL won. The battle made the front page of the *New York Times*. The public was beginning to care about code. This wasn't the code that would crack the secrets of mathematics or the nature of the universe. This code was the lifeblood of our economy and society. It was the summer of 1999, and people were starting to realize that the internet and the web were becoming a new, dynamic circulatory system for information, coordination, and life itself.

The summer before my final year of college, I interned as a software engineer at Microsoft in Redmond, Washington. I was assigned to the group that was building MSN Messenger Service, Microsoft's instant messaging app, and after receiving my degree, I went on to join them full-time. The unwieldy name was cooked up by Microsoft's marketing department, which had a reputation for always picking the clunkiest and least imaginative product names. Buddy List? C U C Me? MSN Messenger? No, it was MSN Messenger Service. I'll call it Messenger for short.

At the time the big players in instant messaging were AOL Instant Messenger (AIM), Yahoo, and ICQ. AIM had tens of millions of users; AOL had become the country's biggest dial-up provider in the mid-nineties by blitzing mailboxes with CD-ROMs, and all AOL users automatically became AIM users. Yahoo and ICQ each had millions of users. These are paltry numbers by the standards of 2018, but they were meaningful in the nineties.

The project was divided into two pieces. On the client team, where I started, we wrote the program that installed on users' computers. We created a minimalist user interface to let people see their buddies when they came online, popping up chat windows for messaging with them, and so on and so forth. We were part of the long-standing Microsoft tradition of creating a product that would do everything our competitors could, with a few new wrinkles added. (This was termed a "best-in-class" product.) The server team was responsible for building the program that tracked all the clients connecting to it and delivered notifications to each user's buddies. If your buddy Gordon logged on, the server would tell your client that he was there. We on the client side had to take the notification from the server and display it properly to the user. The server side also had to integrate our functionality with Hotmail, which had tens of millions of users and which Microsoft had acquired in 1997. It was imperative that every Hotmail user be able to log on to Messenger with a Hotmail address and password as seamlessly as possible. This was not simple. Hotmail and Messenger were two entirely separate systems.

My initial team consisted of about ten people, though it gradually expanded to several times that size. On the client side, we'd meet to discuss what needed to be done, what kinds of features we wanted, what we could do and couldn't do. Then we'd go and do it. As I was the youngest person on the team, I was given little chunks of the project to work on at first, supervised by a meticulous and sharp mentor, Chris. We would go through my code line by line as I took notes, and he would point out my mistakes. I then fixed them, and we would repeat the process. It was very much a guild apprenticeship. I built the instant messaging windows: the "type your message here" window and the "transcript" window above it. I added better font control and helped make the client work with non-Latin character sets like Chinese/Japanese/Korean, Indic, Hebrew/Arabic (right-to-left languages were a breeding ground for bugs). I managed when the windows popped up, how they could be moved around, and how scrolling worked in them (scroll bars were also extremely buggy in Windows). Handling shutdown was especially difficult: we made sure the windows closed down neatly and all the program's resources were cleaned up properly without the program crashing. "If it's shutting down anyway, can't we just let it silently crash?" I joked.

After we finished developing the client code, we had some downtime while waiting for the server team to finish their Hotmail integration. We couldn't release without their work. We fixed every bug we could find. One of our problems was getting new users to join Messenger when so many people already used other chat programs. Unlike email, which followed a common format across all programs and servers, these chat networks didn't talk to one another; AOL didn't talk to Yahoo, which didn't talk to ICQ, and none of them, of course, would talk to Messenger. AOL had the largest user base, so we discussed the possibility of adding code to allow Messenger to log in to two servers simultaneously, Microsoft's and AOL's, so that users could see their Messenger and AIM buddies on a single list and talk to AIM buddies via Messenger. We called it "interop."

It wasn't elegant, but it wasn't too complicated either. A client program like Messenger talks to a server using a well-defined protocol, a set of coded instructions sent to and from the server. HTTP (HyperText Transfer Protocol), used to request and transmit web pages, is one of the most common protocols in existence. It is built on top of TCP/IP (Transmission Control Protocol/Internet Protocol), the underlying protocol of the internet itself. Internet companies run servers that speak these and other protocols. Some protocols, like HTTP and TCP/IP, are public. Some are private and undocumented. AIM's protocol, known as OSCAR (for Open System for CommunicAtion in Realtime), was in the latter category. We didn't have the necessary "key" to decode it. Instead, I signed up for an AIM account and watched the AIM client talk to the server using a network monitor, a development tool used to track network communications in and out of a computer. I saw the protocol that AIM was using. My mentor Chris had figured out a large chunk of OSCAR this way, and after he left the team, I picked up his work and finished the job. A sample message looked like this, with the hexadecimal representation of the binary on the left and the ASCII character translation of the binary on the right:

```
2A 02 EE FA 00 B0 00 0E 00 06 00 00 9B 7D BD 28 *...........}.(
33 41 35 36 35 43 38 37 00 03 00 03 00 28 0D 52 3A565C87.....(.R
45 41 4C 52 65 67 72 65 73 73 6F 72 00 00 00 03 EALRegressor....
00 01 00 02 00 10 00 0F 00 04 00 00 00 18 00 03 ................
00 04 3E 4C BE 8C 00 01 00 00 00 05 00 68 00 02 ..>L.........h..
00 08 75 73 2D 61 73 63 69 69 00 03 00 02 65 6E ..us-ascii....en
00 01 00 48 69 2E 2E 20 41 6E 79 62 6F 64 79 3F ...Hi.. Anybody?
```

Much of this is opaque, but "Hi.. Anybody?," which I wrote into my AIM chat box, appears at the very end, surrounded by the protocol packaging for the message. Some of this protocol was always changing, but some was fixed from message to message. Chris and I decoded the messages. Then we took AOL's protocol, packaged up text messages in it, and sent it to the AOL servers. Did AOL notice that there were some odd messages heading their way from Redmond? Probably not. They had tens of millions of users, and we were using their own protocol, after all. I thought this little stunt would be deemed too dubious and excised from Messenger before it shipped. But management liked the feature. On July 22, 1999, Microsoft entered the chat markets with MSN Messenger Service. Our AOL "interop" was in it.

Reception was generally positive. Everything worked, and it had better international support. It integrated with Hotmail, and users could use Messenger to talk to their AOL chat buddies. Our lark was paying off.

No one had warned AOL of our gambit, and they weren't happy with what we termed our "unauthorized interoperability." Quickly, they blocked Messenger from connecting to their servers by varying their protocol messages in ways we had not anticipated. Users who tried to contact their AIM buddies from Messenger would receive a pop-up instant message saying, "Use an authorized AOL client at this link: [web URL]." But as long as Messenger sent *exactly* the same protocol messages to the AOL servers, AOL wouldn't be able to detect that a user was on Messenger. I took the AIM client and meticulously checked for differences in what it was sending, then changed our client to mimic it once again. Messenger users received an upgrade with this new fix. But AOL again caught on and switched their protocol up. We matched their client again and pushed out another upgrade. We went back and forth at least a dozen times. At one point, AOL sneakily excluded users logging on from Microsoft headquarters from their changes, so that we would be unaware that other Messenger users were receiving an error message. After an hour or two of scratching our heads, a team member sick at home notified us that yes, she was getting disconnected.

Microsoft and AOL were both tech titans in 1999, and soon the press got hold of the story. On July 24, the *New York Times* printed: "In Cyberspace, Rivals Skirmish Over Messaging." AOL kept blocking us, wrote the paper of record. "But Microsoft refused to roll over. Late Friday, the software giant said it had revised its MSN Messenger program to circumvent America Online's roadblock. Within hours, America Online answered that challenge with a new block." It was like reading about a wrestling match in which I was the mysterious masked challenger.

I framed the article. *That was me!*

Microsoft Agonistes

> Most software today is very much like an Egyptian pyramid with millions of bricks piled on top of each other, with no structural integrity, but just done by brute force and thousands of slaves.
>
> —ALAN KAY

This was 1999. Two decades after launching MS-DOS, its first operating system, Microsoft was one of the biggest companies in the world. It employed thirty thousand people worldwide. About ten thousand were in Redmond, and two of them were now me and my wife. The campus was about the same size as Yale's, but moving there felt like being flung into a different world, where technical strengths were deployed in headlong pursuit of profit. Despite Microsoft's notorious "stack-ranking" of employees, a zero-sum game that pitted team members against one another to compete for bonuses, it was still less competitive than academia, with a great sense of camaraderie and teamwork.

I came to Microsoft in its middle age, when it had reached the summit of its unprecedented software dominance. What was Microsoft's secret? They were, and are, a software company. While hobbyists in the 1970s were trying to figure out how to build a computer small enough to fit in your home, Bill Gates and his partner Paul Allen were writing software for that as yet unbuilt machine. In 1980, they partnered with IBM to provide MS-DOS (Microsoft Disk Operating System) for IBM's first mass-manufactured personal computer, the PC. Gates and Allen foresaw that PCs were going to be ubiquitous and that software, rather than hardware, would be the profit point.

By licensing Microsoft to provide the operating system for their PCs, IBM essentially handed them a license to print money. The margins on software were far greater than those on hardware, because the physical manufacturing process was negligible—producing disks was cheap and trivial next to microprocessors and peripherals. And Microsoft had a lock on guaranteed sales of the ballooning PC industry. IBM wasn't the only hardware maker in town—far from it—but Microsoft was the only maker of MS-DOS.

I didn't grasp Microsoft's prominence growing up. They had no presence on the Apple II, which was not a business platform, and their PC software was, at the time, runner-up in competition to more established players like WordStar and Lotus. I understood that they provided the crucial though unflashy operating system to IBM PCs. What I didn't understand was just how tightly and brilliantly they had placed themselves at the core of PC software, with strategies that would enable them to dominate the software business in the 1990s.

Microsoft's rise did not go unnoticed or uncontested. In 1984, Apple debuted the Macintosh. Following the far less popular Lisa, which came out the year before and cost $10,000, Lisa's "little brother" Mac was the second PC to use an operating system with a graphical user interface (GUI), building on research done at Xerox PARC and elsewhere. Apple bought ad time during the Super Bowl to trumpet this revolution in computing, and they weren't exaggerating. Until the Macintosh, PC screens displayed only text and the most primitive of graphics, but now Macintoshes featured a visual representation of the inside of the computer—a "metaphorical desktop," as it was called. I found it dazzling, but computers weren't yet powerful enough to make the GUI indispensable. Toward the end of the eighties, home computers became fast enough to make multitasking increasingly valuable, and it was clear that GUIs promised far more user-friendliness than text command lines.

Microsoft thought so too. They partnered with Apple to provide software to Mac users, including Microsoft Word. More significantly, in 1985 they released the first iteration of Windows (with, importantly, some GUI elements licensed from Apple, such as icons, windows, and scroll bars). Windows featured a clickable list of the files on the computer, resembling today's Windows Explorer, plus some other "windows" displaying executable files (a calculator, for example). It was an improvement over the MS-DOS command prompt, but a far cry from the Macintosh's elegant monochrome desktop. In 1987, Microsoft released Windows 2.0. This was still clunky, but already better, with overlapping windows and more logical functionality. Apple could see where things were headed, and in 1988 they sued Microsoft for copyright infringement.

The suit failed. Windows was similar to the Mac operating system, but hardly identical. The appeals court wrote, "Almost all the similarities spring either from the license [for the initial Windows] or from basic ideas and their obvious expression. . . . Illicit copying could occur only if the works as a whole are virtually identical."

The initial decision came down in 1992 and was affirmed on appeal in 1994. It was a serious blow to Apple, already in its Steve Jobs–less slump. Hampered by poor management, overpriced computers, and the protectionist attitude that only Apple could make Macintosh hardware, the company saw its market share decline throughout the decade, eventually prompting the return of the exiled Jobs and the company's resurgence. Windows, meanwhile, conquered the world. In Windows 95, the first post-lawsuit release of the operating system, Microsoft incorporated Apple's famous trash can, impishly refashioned as a "recycle bin." For a good long while, Windows could not be stopped. It never reached the elegance of MacOS, but it didn't matter. Windows was everywhere.

Gates and Allen were skilled and even brilliant coders, but the history of software is littered with people just as smart or smarter who did not end up billionaires. Their true achievement was on the business side. For years Microsoft remained a small company. In the 1980s and '90s, companies didn't need to be big to make software. The programs were less complex, and there wasn't yet so much competition, so Windows could charge a premium for them. The amount of person-hours that goes into a $50 piece of software today dwarfs that for a $50 item of software thirty years ago—one of the reasons why the software industry doesn't mint millionaires like it used to. In 1983, the prestigious word processor Word Star was so primitive it required users to put little stickers on their keyboards so they'd know which functions correlated to which keys. This program retailed for $289.

In this world, Microsoft stood out. They worked fast, they were aggressive, and they were very cagey. Their strength was never in innovation per se, but in appropriation, improvement, and integration. One slogan that I would hear bandied about at the company was that Microsoft made "best-in-class" products. A less charitable way to put this would be to say that Microsoft would develop products that were better *enough* than the best out there to take over the market. The quality of Microsoft's offerings closely tracked the quality of existing offerings.

Lotus's spreadsheet software 1-2-3 was a solid product in the 1980s and early 1990s; consequently, Microsoft Excel, which debuted in 1985, became the standout of Microsoft's nascent Office suite. Word processors like WordPerfect and WordStar were less formidable; as a result, Microsoft Word was considerably less robust than Excel. And in the absence of any dominant email programs, Microsoft Outlook was buggy and slow. It remained that way well into the early 2000s—much to the vocal dismay of Microsoft employees themselves, who were stuck using it.* Microsoft was far too efficient to waste time improving a project beyond what was needed to defeat their competitors. In the late nineties I got a chance to tour the legendary Massachusetts computer company Digital Equipment Corporation (DEC, later bought by Compaq), and the difference in culture was remarkable. There were people at DEC who had been working on threading (the manner in which

* This process is called "dogfooding," referring to a company that claimed its dogfood was so good its own employees ate it.

operating systems manage concurrent sets of processor instructions) for *twenty years*. Half the people had PhDs in their areas of specialty. Corners were never cut for a faster release. Ah, I thought, this is why Microsoft won out.

Microsoft certainly *tried* to innovate with new products from time to time. Clippy, the little paper clip that popped up occasionally in Microsoft Word, was an innovation. Microsoft Bob, a yellow dog in dark glasses who showed up in Windows 95 to offer help, was an innovation. Cairo, Microsoft's "revolutionary" new operating system from the 1990s, would have been a tremendous innovation had it ever shipped. But Microsoft was far more comfortable entering existing markets and besting competitors. In the absence of a clear target, planning became fuzzy and tentative. This was apparent in their reticence to engage wholeheartedly with the internet in the 1990s. No one was making gobs of money online yet, so Microsoft didn't have anyone to emulate. It wasn't as if Microsoft didn't realize that there was money to be made. Microsoft just wasn't about to create the mechanism to do so on its own.

By 1999, the Windows and Office behemoths had ensured Microsoft's ongoing dominance of the desktop operating system and business applications markets for as long as the PC remained central to businesses and consumers—even when the *US v. Microsoft* antitrust trial was at its peak in 1999–2000. The suggested plan to split Microsoft into two monopolies, one for Windows and one for Office, wouldn't have helped, even if it had made it past the appellate court that overturned the initial judge's ruling. The case was a bizarre, political sideshow, which had only a minor impact on the tech industry. It did ensure that future tech companies kept a far larger battery of lawyers and lobbyists close by. The internet—and Microsoft's inability to engage with it productively or profitably—hurt Microsoft far more than the antitrust suit did.

Through the antitrust discovery process, I learned that before I arrived, there had been a war over the future of the company at the highest levels of Microsoft—between the "doves" and the "hawks." The doves wanted to make common cause with other internet companies, like Netscape and AOL, and share power with them. The hawks wanted Microsoft to be the exclusive provider of internet services. The real bone of contention was the future of Windows: here was the most profitable thing in the history of computers. It seemed foolhardy to sacrifice the centrality of Windows to the open internet, but a truly aggressive internet strategy would have meant thinking about a world where Windows was *not* on every computer and device. "I don't want to be remembered as the guy who destroyed one of the most amazing businesses in history," one senior executive wrote of Windows during this argument. In the end the hawks won and most of the doves left Microsoft. Then the hawks lost.

Code Work

> Anyone who has had the misfortune to write his first computer program remembers the humiliation in conversing with a servant or master that insists on a language unworthy of the dullest of intelligences and the lowest of men.
>
> —WAN-LEE YIN

At Microsoft, as at Google, I worked in C++, a language that exudes compromises between efficiency and elegance. They are often ugly compromises, but in software engineering, an ugly compromise trumps beautiful purity every time.

C++ is a significant step up from assembly, but considerably closer to the hardware than most other commonly used languages (including BASIC and Logo), which means that C++ is hard to beat for sheer speed—crucial when you're writing a server and your code has to execute on thousands of machines simultaneously. A 10 percent speedup can mean a 10 percent decrease in capital expenditure.

C++ is a "mid-level language," in between the low-level language of assembly and high-level languages like Java and Python. The higher the level of the language, the less control it provides. Assembly lets a programmer specify exactly where in the CPU every bit of information is going. Higher-level languages offer less control over the guts of the computer, which is managed by compilers, interpreters, and virtual machines. These programs are exceedingly good at managing things automatically. They don't make mistakes (unlike humans), but they do have their limits. In particular, they cannot understand the overall intent of a program, like an automatic transmission, which can't quite match the manual transmission for performance because it can't anticipate the driver's intent. If you pile on too many abstraction layers and automated tools, performance suffers. The downfall of Microsoft's notorious Vista operating system came about because the majority of the code was written in a new language of Microsoft's own design, called C#. Like Java, C# was considerably higher level than C or C++, and the code responsible for taking care of the lower-level nastiness didn't perform well enough. So Microsoft had to scrap the C# code and started over in C++, costing them an extra two years of work. Lesson learned.

Any computer language like C++, Java, or Python consists of a certain number of commands (like **`repeat`** in Logo or **`GOTO`** in BASIC), often not more than a few dozen, and a certain number of numerical and logical operators, like **+** and **&&**. Many languages offer similar command sets; where they differ is in the methods they provide for structuring programs and in the amount of abstraction they provide from the underlying computer fundamentals.

Early on in my time at Microsoft, the code was all that mattered to me. The Messenger client was the first piece of shipped, professional code I worked on. The Messenger client at Microsoft at the time was about a hundred thousand lines of code, all in C++. These implemented the display of the user's list of contacts, online and offline; pop-up notifications when a buddy logged in; uninstallation code to remove the program if people hated it; IM archival functions; connections to the Messenger servers (and, for a while, the AOL servers); and more. In its early versions, it was an efficient, lean little program.

I no longer have access to the Messenger code, which remains the private intellectual property of Microsoft. Instead, here is a piece of the open-source C code for the chat program Pidgin. This function, **`update_typing_icon`**, is called to update the "typing indicator," which pops up to show when a buddy is typing out a message. I built this feature into the Messenger client, and I still have a fondness for it.

```
static void
update_typing_icon(PidginConversation *gtkconv)
{
    PurpleConvIm *im = NULL;
    PurpleConversation *conv = gtkconv->active_conv;
    char *message = NULL;
    if (purple_conversation_get_type(conv) == PURPLE_CONV_TYPE_IM)
        im = PURPLE_CONV_IM(conv);
    if (im == NULL)
        return;
    if (purple_conv_im_get_typing_state(im) == PURPLE_NOT_TYPING) {
        update_typing_message(gtkconv, NULL);
        return;
    }
    if (purple_conv_im_get_typing_state(im) == PURPLE_TYPING) {
        message = g_strdup_printf(_("\n%s is typing..."),
            purple_conversation_get_title(conv));
    } else {
        assert(purple_conv_im_get_typing_state(im) == PURPLE_TYPED);
        message = g_strdup_printf(_("\n%s has stopped typing"),
            purple_conversation_get_title(conv));
    }
    update_typing_message(gtkconv, message);
    g_free(message);
}
```

The function takes a parameter called **gtkconv** that contains information about the chat session (**PidginConversation**) being updated. It calls another function called **purple_conv_im_get_typing_state**, passing it to the chat session in question. That function then returns one of three possible values: **PURPLE_NOT_TYPING**, **PURPLE_TYPING**, or **PURPLE_TYPED**. ("Purple" is the name of the core chat library interface. Coders use silly names.) A user interface function, **update_typing_message**, then changes what message is displayed on the screen. In the case of **PURPLE_TYPING**, a message with "[Buddy name] is typing" is shown. If **PURPLE_TYPED**, meaning that text has been entered but your buddy hasn't typed anything for a bit, "[Buddy name] has stopped typing" is shown. And if no text has been entered and the buddy isn't typing (i.e., **PURPLE_NOT_TYPING**), then no message is shown at all.

update_typing_icon calls other functions like **update_typing_message** and **purple_conversation_get_title**. Most of these other functions are also part of the Pidgin program, separated into modular chunks so that each can be isolated, tested, and perhaps reused. One exception is the **g_strdup_printf** function, which creates the text string containing the message to be displayed. **g_strdup_printf** is part of the open-source GNOME user interface library. It is sufficiently generic to be of use to many programs, not just Pidgin.

All this C code is compiled into assembly by a C compiler. Microsoft had its own, marketed as a part of Visual C++, while there also exists the popular, free gcc compiler, which I used at Google. The resulting executable file of assembly code can then run natively on the processor for which the compiler was designed. And there you have yourself a chat client.

The Buffer Overflow

Progress is the exploration of our own error.

—JACOB BRONOWSKI

The Messenger war was a rush. Coming in to work each morning to see whether the client still worked with AOL was thrilling. If it wasn't, I'd have to look through reams of protocol messages to figure out what had changed, fix the client, and try to get an update out the same day. I felt that I was in an Olympic showdown. I had no idea who my adversaries were, but I had been challenged and *I wanted to win*. Our users cared too. They wanted us to win.

AOL tried different tactics. At one point I suspected that they identified the Microsoft client because it wasn't downloading the advertising that the AOL client downloaded. I updated our client to download it all and then throw it away. AOL included mysterious protocol messages that didn't seem to affect their client but broke ours. I fixed that. One day, I came in to see this embedded in the messages from the AOL server:

"HI. —MARK." It was a little missive from engineer to engineer, hidden from the corporate, media, and PR worlds that were arguing over us. I felt solidarity with him even though we were on opposing sides.

AOL was publishing propaganda about how Microsoft was behaving like an evil hacker by asking for AOL passwords. This wasn't true, but we were allowed to respond only through our PR department. My team was sealed off—but our code wasn't.

Then AOL stopped blocking us. It was strange to encounter sudden silence, as though the enemy had abruptly yielded the battlefield, and while I wanted to believe we'd won, I suspected AOL wouldn't give up without a word.

A week later, we found that Messenger had been blocked again, but this time was different. The AOL server was sending a huge chunk of new gobbledygook that I could not understand. It looked approximately like this:

```
                                2A 02 77 9C 01 28 00 01          *.w..(..
00 13 00 00 80 0E A6 1B 00 FF 00 0B 01 18 83 C4 ................
10 4F 8D 94 24 E4 FE FF FF 8B EC 03 AA F8 00 00 .O..$...........
00 90 90 90 90 8B 82 F0 00 00 00 8B 00 89 82 4E ...............N
00 00 00 8B 4D 04 03 8A F4 00 00 00 8D 82 42 00 ....M.........B.
00 00 89 45 10 B8 10 00 00 00 89 45 0C C9 FF E1 ...E.......E....
00 01 00 20 00 00 00 00 00 00 04 00 00 00 00 00 ................
00 00 00 00 00 00 00 00 00 00 00 00 00 00 00 00 ................
00 00 00 00 00 00 00 00 00 00 00 00 00 00 00 00 ................
00 00 00 00 00 00 00 00 00 00 00 00 00 00 00 00 ................
00 00 00 00 00 00 00 00 00 00 00 00 00 00 00 00 ................
00 00 00 00 00 00 00 00 00 00 00 00 00 00 00 00 ................
00 00 00 00 00 00 00 00 00 00 00 00 00 00 00 00 ................
00 00 00 00 00 00 00 00 00 00 00 00 00 00 00 00 ................
00 00 00 00 00 00 00 00 00 00 00 00 00 00 00 00 ................
00 00 00 00 00 00 00 00 00 00 00 00 00 00 00 00 ................
00 00 00 00 00 00 00 00 00 00 00 00 00 00 19 10 ................
08 11 29 EC FF FF 44 00 00 00 00 00 00 00 FF 00 ..)...D.........
00 00 08 01 00 00 00 00 00 00 90 47 40 00 F8 E9 ...........G@...
EA FE FF 00 00                                  .....
```

The first couple of lines are the standard AOL instant message protocol header. But from **90 90 90 90** onward, it's incomprehensible, bearing no relation to anything the AOL servers had sent their client or ours. The vast expanse of double zeros in the middle was also mysterious, since a bunch of zeros couldn't contain much meaning.

Our client ignored it, but the AOL client responded to this gobbledygook with a shorter version of the same gobbledygook. I didn't know what it was. It was maddening. After staring at it for half a day, I went over to Jonathan, a brilliant server engineer on our team, and asked what he thought. He looked at it for a few minutes and said, "This is code." It was actual x86 assembly code. The repeated 90s tipped him off: they signify a "no-op" no-operation instruction in x86 assembler, telling the processor to do nothing for one cycle.

The pieces then came together. Normally, protocol messages sent from the server to the client are read as data, not as code. But AOL's client had a security bug in it, called a buffer overflow. A buffer is a place in memory where a program temporarily stores data during execution. It's all too easy in lower-level languages to allow in more input than the buffer can accommodate. In this case, large protocol messages flooded over the end of the buffer. In the computer's memory, the locations past the end of the buffer are often filled with things that aren't just stored data. The actual assembly code to be executed is often nearby. This buffer overrun could overwrite the client code currently being executed and control the functioning of AOL's program.

It's a huge security hole, since it gives the server control of the client PC. In the wrong hands, the server can shut down or spy on a computer. AOL knew about this bug in their program, and now they were exploiting it! The double zeros were filling up space in the program's buffer so that once the large message hit the end of the AOL client's buffer, it would overwrite executable code with the remainder of the protocol message. The remainder of the protocol message contained *new* code to be executed, which the client promptly did. AOL caused the users' client to look up a particular address in memory and send it back to the server.

This was tricky—vastly trickier than anything they'd done so far. It was also a bit outside the realm of fair play: hacking into *their own client* using an unfixed security hole—a hole our client didn't possess. A "Rommel, you magnificent bastard" moment. I was out of my depth.

Someone else at Microsoft—I never found out who—told the press about the buffer overflow, figuring that if people knew that AOL's client had a huge security hole in it, AOL would be forced to patch their client and would no longer be able to exploit it.

According to security expert Richard M. Smith, a certain "Phil Bucking" of "Bucking Consulting" sent him a message, alerting him to the buffer overflow in the AOL client:

> Mr. Smith,
>
> I am a developer who has been working on a revolutionary new instant messaging client that should be released later this year. Because of that, I have followed with interest the battle between AOL and Microsoft and have been trying to understand exactly what AOL is doing to block MS and how MS is getting around the blocks, etc. Up until very recently, it's been pretty standard stuff, but now I fear AOL has gone too far.
>
> It appears that the AIM client has a buffer overflow bug. By itself this might not be the end of the world, as MS surely has had its share. But AOL is now *exploiting their own buffer overflow bug* to help in its efforts to block MS Instant Messenger.

Getting the name of MSN Messenger Service wrong was a nice touch, but the rest of it is rather ham-fisted. This developer of a revolutionary new app takes sides in the Microsoft-AOL war without promoting his own app? The email also includes a trace of the buffer overflow message itself, which I remember vividly from the hours I spent staring at it. If Phil Bucking's text—and his name—weren't suspicious enough, he'd also sent the message (via a Yahoo account) from one of Microsoft's computers at a Microsoft IP address. The IP address showed up in the email headers. Microsoft's digital fingerprints were all over the email.

Smith accused Microsoft of sending the email. Microsoft fessed up. The news story that emerged covered Microsoft's attempt to bad-mouth AOL under a fake identity—an easier sell than explaining the buffer overflow. People on various security forums ascertained that the buffer overflow was real and inveighed further against AOL, but the press wasn't paying attention. The buffer overflow persisted into several later versions of AOL's client. AOL never admitted a thing, and the press never did understand it.

We gave up. I licked my wounds. I switched to the server team, which I preferred to the client, but working with MSN Messenger Service gradually became dreary and politicized. Those were the years of Microsoft's long, slow decline (which reversed only when Steve Ballmer was removed and replaced by Satya Nadella as CEO in 2014). Bureaucracy and overhead ballooned. Getting approval for any idea required running a gauntlet of meetings with management. The infamous "stack rank" review system pitted teams and individual engineers against one another. There was an incredible thirst for "headcount" within divisions, so managers would lobby aggressively to move independent groups under their control. The ambitious and forward-looking NetDocs, an internet-based document-editing suite, gobbled up a number of small groups in the late nineties. But then NetDocs got eaten by Office, which killed NetDocs before it released. The market was open when Google debuted its own online word processor, Google Docs, in the mid-2000s. And so it went. Multiyear projects with hundreds of engineers died without the public ever hearing a word.

On September 11, 2001, my wife was away on a trip. She woke me up early with a phone call to tell me about the World Trade Center attacks. I struggled to process the news with my foggy brain, then drove to work. The televisions in the lobby were broadcasting MSNBC's reporting. A few people stood in front of them, but no one was talking about it. Employees were quieter than usual, but the endless parade of meetings otherwise continued unchanged. It was the first hint I had that the heads-down mentality at Microsoft was removing its employees from other important aspects of the world. Complacent from success, Microsoft had turned inward and was attacking itself in much the same way one's immune system can attack healthy cells in the absence of external threats. The ecology had gone wrong.

During the bad days, I got to see how average people performed under conditions that encouraged them to behave unethically and dishonestly. Some stuck to their principles, trying to do good work on the hope that rationality and good arguments would prevail over groupthink and nepotism. Others caved, sacrificing their integrity in order to line up behind some manager's empty rhetoric or secretly playing both sides in political disputes. I saw old boys protecting and boosting their own. I saw cowardice and deceit. When my team opposed the impossible ambitions of an ambitious project manager, he responded by creating another, hand-picked team to take over our work and chase his dreams. (It didn't work out.)

For my part, I didn't betray my principles, but I failed to defend them against corporate inertia. My greatest regret is toward the people I managed, whom I was not able to protect from corporate caprice. Like many low-level managers, I had more responsibility than actual authority, leaving me the deliverer of bad news I couldn't control. It's a dilemma I have not managed to answer to this day. At the age of twenty-four, I had no idea at all. My direct reports, all of them good people, got lower review scores than they deserved, and I told them that they were doing well even as I wanted to tell them that the system had gone rotten. The stress turned me brittle and harsh. I gave others too hard a time, to say nothing of myself.

I endured through the friendship of a couple others on my team. We reassured one another that it was the system that was crazy, not us. We were right; we all became much happier after we left. The personal and professional wounds of corporate life may not sound dramatic, but the power an employer has to define a person's identity should not be underestimated. When I left Microsoft and went to Google, corporate paranoia lingered over me for months, until the healthier and less cut-throat environment gradually subsumed my emotional miasma.

After a year or two of ongoing malaise, bitter jokes, and festering anxiety, I quit Microsoft for a job at Google. My boss threatened me with litigation for leaving. I packed up my things in a box and walked out the same day. I heard similar stories from other engineers who left for Google. When Windows architect Mark Lucovsky took a job at Google, Steve Ballmer allegedly threw a chair across the room (or at least shook it). Microsoft was hemorrhaging hundreds of top engineers to Google, and the combination of the talent loss plus the insult to the executives' egos made for very bad blood.

Despite the toxic politics, MSN Messenger Service did pretty well. Messenger acquired tens of millions of users. Millions used Messenger at any one time. I added emoticons in 2000—it was the first American chat program to turn a colon and a close-parenthesis into a smiley face (the South Koreans may have preceded us)—and people loved it. We added internet phone calls to the client. I helped redesign the server architecture, which was the greatest technical challenge I'd faced. I was glad to have worked on it.

After I left Microsoft, Messenger puttered along for years. When Microsoft's purchase of Skype rendered it both redundant and obsolete, Messenger was finally retired in 2014. By that point, I hadn't used it in years.

The larger lesson of the Microsoft-AOL war was that the workings of code were no longer a private matter known only to those steeped in the lore of computing. They were increasingly becoming public affairs that impacted national and even global events. The chat wars were, to me, a duel between a handful of people within two gargantuan companies, and yet they had become a national news story for a brief period. Looking back, it was the first inkling I had that it wasn't just *computers* that were permeating all our lives, but the *code* itself. Yet few grasped code's function and its impact—something that is still true today. The ultimate outcome of the conflict resulted from a buffer overflow whose existence was never agreed upon, because neither reporters nor the public understood it.

Some time after I left Microsoft, I met one of the AOL engineers who'd worked against me during the interop war. We had a huge laugh over it. I complimented him on the genius of the buffer-overrun exploit, even as I bemoaned my loss. "It had been a great game," I said. He agreed. We were both dumbstruck that we were still two of a very small number of people who knew the real story, after all these years.

The other lesson of the chat wars is that one cannot understand the technological or sociological impact of software without a solid grounding in code. That understanding has to begin from the bottom up, from the fundamental essence of code and computation: the binary bit.

3

BINARIES

Praxis and Theory

> Our ordinary habits make it easy to talk as though in saying "computing" we were dealing with something well delimited, an item of a sort. . . . The matter is more subtle than that. Instead I consider computing to be a human activity involving certain human purposes and intents, certain human insights, and certain man-made tools and techniques.
>
> —PETER NAUR

AT MICROSOFT, I dealt with buddy lists and chat sessions, but it didn't matter to me who these people were or what they were saying. I cared only about shuttling notifications and delivering messages, like the postal service. Like the computers I programmed, I was disconnected from any human meaning of the data. There were only the generic forms of "people" and "messages." Otherwise, the data was as opaque to me as it was to the servers. Both I and my programs were concerned with computation, not understanding.

Computers are first and foremost machines for doing calculations far more quickly and perfectly than humans. It's even there in the name: the word "computation" historically referred to the reckoning of numbers. When Milton writes of the constellations' "starry dance in numbers that compute / Days, months, and years," he is describing the heavens' computation of numerical time. Yet there's very little in Logo, instant messenger, and the buffer overflow that appears to be mathematics. The computer is indeed doing math—simple math, for the most part. But computation isn't arithmetic so much as *the act of doing arithmetic,* and the significance of computers' work today is that the numbers they compute represent our reality. They represent it very crudely, and they do so with great trouble, but we use computers by making their numbers mean something to us.

The history of computers developed along two parallel paths. On the one hand, there is the engineering work of building better machines that do calculations. On the other, there is the theoretical work of exploring what those calculations can *mean* algorithmically and in our world. In the 1930s and 1940s, Alan Turing theorized different models of computation, while engineers like Konrad Zuse (who built what was arguably the first programmable computer, the Z3, in 1941) created increasingly general-purpose computing machines with scant knowledge of their theoretical possibilities and limitations. Computer science (the theory) and software engineering (the practice) influence each other bidirectionally, and both sides have been critical to the evolution of computers as we know them. Theoreticians gave us more sophisticated computer languages to program in. Theoreticians identified the artificial intelligence models that allowed for supervised training on datasets. But it was only when deployed on the huge datasets and clusters at Google that these models yielded revolutionary results, in areas such as image recognition and Go strategy. Many, including myself, were shocked when Google's AlphaGo beat top human players at Go in 2016. The sheer combinatorial complexity of Go had me convinced that it would be another decade before a computer could take the top spot. The maxim known as Moore's Law (after Intel co-founder Gordon Moore) states that the number of transistors that engineers are able to fit into an integrated circuit doubles approximately every two years. In other words, raw computational power increases exponentially. The top computers contained thousands of transistors in 1970; today they have billions. But computers are not necessarily millions of times smarter than they were then, and their achievements, like AlphaGo, can still come as surprises to us. Machine intelligence depends on computing power, but only loosely. It also depends on intellectual progress in computer science.

There are computer scientists who make a point to do as little programming as possible. There are many coders who have only a passing familiarity with the theory of computer science. As with most divides, each side would benefit from more familiarity with the other. Machine learning, in particular, has greatly profited from the close relationships between engineers and scientists, because the practical results of machine learning are often strong indicators of what approaches to pursue. Yet a theoretician's eye is still needed to question machine learning paradigms and construct new approaches. I chose to become an engineer because I wanted to *build* things rather than *prove* them, but without the theoretical structure provided by my background in computer science, I wouldn't have been half the engineer I was.

Computers today are a thick layer cake of abstractions. Many engineers work at one level while being only vaguely conscious of the other. The chip designer doesn't think about web pages. The web designer knows little of silicon. But every serious software engineer understands how the layers fit together, and just as importantly, how they tend to recapitulate similar theoretical ideas of structure and purpose. And that knowledge then informs engineers' design of *new* abstraction layers, making them better, more robust, and more elegant.

Movies and television present code as cryptic lines of arcane commands on a screen. This image is a mystifying ruse. A programmer should be able to look at good code and, with some effort, derive a fundamental *simplicity* to its organization and function—a set of organizing principles that structure the code in a neat and predictable way. This understanding arises from one's knowledge of the hidden reality that lies behind all code. And that reality, in a word, is logic.

Truthtellers and Liars

> All that logic warrants is a hope, and not a belief.
>
> —CHARLES SANDERS PEIRCE

There is a type of puzzle, made most famous by the logician Raymond Smullyan (and by the movie *Labyrinth*), that involves liars and truthtellers. Suppose there are only two types of people, those who *always* tell the truth and those who *always* lie. Assuming that, can we determine who is the liar and who is the truthteller in this exchange?

MOLLY: Leopold is a liar.
LEOPOLD: We are both liars.

The reasoning goes that Leopold cannot be telling the truth, because if he is a truthteller, his statement is a lie, which would be a contradiction. So he must be a liar. Therefore, his statement is false and Leopold and Molly are *not* both liars. Since Leopold is a liar, Molly must be a truthteller. And sure enough, Molly's statement is true.

Here's a variant that I gave to my six-year-old daughter. There are two doors. Behind one is a huge amount of money. Behind the other is a shark. You don't know which. There's writing on the doors:

DOOR 1: The shark is in here!
DOOR 2: Both doors are lying.

Which door do you open? The logic is the same as with Molly and Leopold. Door 2 must be lying, because it's impossible to truthfully say that you are a liar. So Door 1 must be true (or else Door 2 would be telling the truth), the shark is behind Door 1, and the money is behind Door 2.

I compulsively worked through hundreds of these sorts of logic puzzles as a kid. I'm fortunate my mother stumbled upon Smullyan's *Alice in Puzzle-Land*, which became my introduction to the field of Boolean logic (the binary logic that underpins all computers), liar and truthteller puzzles, and logical paradoxes. Smullyan's use of Lewis Carroll's heroine aptly suited his truthteller puzzles, as the rigid yet arbitrary rules of logic matched up perfectly with the original Alice books. Carroll had been an expert logician himself.

After *Alice*, I located Smullyan's other books, notably *What Is the Name of This Book?* and *The Lady or the Tiger?*, which presented similar puzzles in disparate settings like Transylvania and Renaissance Italy. His other books delved into Taoist and Zen-inflected parables but also discussed analytic philosophy, to which Smullyan had contributed as a logician.

Some of the puzzles in these books grew extraordinarily ornate, requiring multiple levels of inference. Sometimes I wasn't told what one or another person had said. Sometimes "yes" and "no" were replaced with nonsense words that meant either "yes" or "no," but I wasn't told which. There were abstruse puzzles that attempted to explain Kurt Gödel's incompleteness theorem—unsuccessfully, in my case.* But one puzzle genuinely irritated me. This is a simplified version:

A man is trying to figure out which of two caskets contains a portrait. "THE PORTRAIT IS NOT IN HERE" is written on the first casket. On the second casket: "EXACTLY ONE OF THESE CASKETS IS TELLING THE TRUTH." Where's the portrait? I went through the possibilities. Whether or not the second casket is telling the truth, the first casket must not be: either the second casket is true and the first casket is therefore false, or the second casket is false and thus the first casket must be false as well (or else the second casket would be telling the truth!). In the puzzle, the man also figures the portrait is in the first casket. He opens it up, and finds . . . nothing. The portrait is in the second casket. He's utterly baffled. I was baffled too. The answer seemed indisputable.

Here is what Smullyan wrote as an explanation:

> Without any information given about the truth or falsity of any of the sentences, nor any information given about the relation of their truth-values, the sentences could say anything, and the object could be anywhere. Good heavens, I can take any number of caskets that I please and put an object in one of them and then write any inscriptions at all on the lids; these sentences won't convey any information.

I felt cheated. I felt that Smullyan had made an implicit promise that he would be providing logic problems that made sense! But, I grudgingly observed, the other puzzles explicitly claimed that the rules were being obeyed, and in this puzzle, there was no such guarantee. The lesson was that truth and falsity are not absolute, but relative to a context.

* Gödel's incompleteness theorem proves, among other things, that certain computer programming problems *cannot* be solved, as the mere existence of their solutions would lead to a logical contradiction.

Outside that context, statements lose sense and meaning. Without the guarantee that the rules are being followed, we can never be sure that our reasoning is valid. Nowhere is that more true than in the logical, and contextually isolated, world of computers.

1s and 0s

> What is a man so made that he can understand number and what is number so made that a man can understand it?
>
> —WARREN McCULLOCH

The mark of the computer is the mark of the digital: everything is encoded in binary, 1s and 0s. The unit of the computer, and of data itself, is the bit, a single binary digit. All digital data, no matter what it appears to be, is at its heart grounded in its core representation in a language of two symbols, or alternatively, a language of one symbol (1) or the absence of a symbol (0). A language of exactly two possibilities.

On the surface, our world does not conform to this language of 1s and 0s. So computational operations, which are built on top of binary, appear alien. Plenty of things, however, can be represented on the computer, and therefore in binary: text, graphics, sounds, videos, programs. As the phenomena and their representations become more complex, some people argue that the digital translation is a corruption of the original.* The argument has been made against CDs, digital video, even

* So it is with audio, where audiophiles insist that no digital process can quite replicate the experience of listening to the analog performance of a vinyl record channeled through a diamond-tipped stylus on a turntable, preferably amplified through transistor-free vacuum tubes. I do not buy these claims. Whatever differences there are (and there surely are some) between records and digital playback, they are capable of being captured within a digital representation. The fetishistic denial of the ability to translate frail vinyl into the robust realm of the digital gets turned on its head when audiophiles try to "shield" digital cables from interference and corruption, as though the error-checked series of bits flowing through them could actually become corrupted (they can't). There is a market for "high-performance" CDs as well, which somehow serve up the exact same sequence of 1s and 0s in a mysteriously enhanced fashion. Audiophiles want to be convinced that they are experiencing *more* and *truer* music than the ordinary listener. In Hong Kong, audiophiles hang violins and cellos from the walls and ceilings in meticulous but incomprehensible patterns, their resonances (or auras) supposedly enhancing the quality of music played in the room.

digital watches. The "aura" of the flesh-and-blood original is lost, or some ineffable vital essence is corrupted and removed in the process. But this is not the case. What we think of as getting lost in the analog to digital translation is actually what is lost in *any* inexact translation from one form into another. A cassette dub of a vinyl record loses more than ripping a CD to a digital file.

Theoretically, anything can be represented in binary—a book, a piece of music, a cellular organism, a star system. In practice, however, we don't yet have the capabilities to do this kind of digital specification for complex structures, so the human brain (about as complex as anything gets) will not become translatable into computable binary representations anytime soon. The process of representation does not necessarily seek perfection, but *transparency*, the point at which the digital replica becomes practically indistinguishable from the original phenomenon. Meteorologists use computer simulations to produce vastly more sophisticated weather models than existed a few decades ago, but weather is so chaotic that we only produce better educated guesses, not perfect ones. On the other hand, acoustic simulations have become so sophisticated that audio technology can reproduce room ambiences and echoes quite precisely, and synthesis of acoustic instruments becomes more sophisticated year by year. The digital recipe does not contain the aural experience of the music any more than a weather simulation creates rain, but a sufficiently precise recipe is what allows for the original phenomenon (music, cells, weather) to be re-created, given the right resources and tools. A computer screen can only show us an image of a painting, but were we to map every speck of paint in that painting, marking every variation in fabric and tone and color, and every daub and splatter, a sufficiently sophisticated machine could re-create the painting exactly—theoretically. In practice, computers and robots could become good *enough* to produce *humanly* indistinguishable copies, given sufficient technological advancement. All the infor-

---

I confess to the inverse obsession. I care about measurable differences in sound that the human ear *can't* detect, instead of unmeasurable differences in sound that the ear supposedly can. I prefer lossless digital formats like FLAC over lossy formats like MP3, even when tests have shown no human ear can tell the difference (certainly not mine, which can no longer hear above 15 kHz due to age). The *knowledge* that the music has been modified to be incomplete—that digital *certainty*—digs at me even though my ears are sadly ignorant of the difference. I store my digital music in pristine, lossless form.

mation one needs to create the picture can be captured. But what is the essence of the original painting? Is it the physical object itself, or the data that describes it?

This is a trick question. The painting *itself* is the data, albeit in analog form rather than digital. By scanning it, we *translate* it into a different, less aesthetically pleasing recipe, yet one that contains the same information. "Things" themselves are *already* representations in one language or another. *Everything* is a representation of some underlying conception of ours. A film is represented in ink on a negative and in photons against a screen. Music is represented in the magnetic organization of oxide on a tape—or, more purely, in the series of waves that resonate against one's eardrum. A chair is represented as a particular configuration of wood and metal; a physical book as patterns of ink on pages. We just tend not to think of these analog physical objects as representations (or recipes) of abstract concepts, but they too contain data. We tend to draw arbitrary distinctions between analog and digital encodings. These distinctions do not hold up. "Data" can come in any number of formats, not just the digital.

Distinctions between reality and the data that represents it are blurred, yet we cling to them. For example, Orthodox Jews are not permitted to erase or desecrate the written name of God. They avoid writing the name of God at all, and so avoid the need to store the papers in a synagogue or bury them in a cemetery.* In the late twentieth century, rabbis needed to assess the issue of writing the name of God on a computer screen. In 1998, Rabbi Moshe Saul Klein ruled that sacred names may be freely erased from a screen or disk. Klein's assistant Yossef Hayad said:

> The letters on a computer screen are an assemblage of pixels, dots of light, what have you. Even when you save it to disk, it's not . . . anything more than a sequence of ones and zeroes.

Letters written on a page, however, are also an assemblage—of particles of ink. Other rabbinical analyses have made a distinction between "permanent" and "temporary" forms of writing, noting that the letters

* This restriction usually applies only to the Hebrew, though you'll frequently see expressions like "G-d" or even "L-rd" in English. Halakhic law applies sacred restrictions to seven names, the most common being the tetragrammaton and Elohim.

on a screen are erased and redrawn dozens of times per second. So the restriction, then, is on a "permanent" assemblage of some kind of "ink" in the physical shape of the name of God. Braille is permanent, for example. Even here there are ambiguities: What would the rabbis think of LED message board arrays? Dry-erase boards? Magnetic sketch boards? Scrabble tiles?

I don't mean to cause more trouble for rabbis. Rather, I want to point out the imprecision of the categorization. We think that objects in our world divide neatly into concrete or virtual, digital or analog, permanent or temporary, and the like. But there are cases that don't quite fit into either category. Orthodox Jews often avoid typing the name of God on a computer to avoid the risk that someone might print out their page and inadvertently create a "permanent" hard copy. While the bit's distinction between 0 and 1 is absolute, it's rare for that distinction to hold up robustly in the complex phenomena of life. It holds up *most of the time,* but that's not quite good enough, as we'll see.

On and Off

> Those who say mathematics is logic are not meaning by "logic" at all the same thing as those who define logic as the analysis and criticism of thought.
>
> —FRANK RAMSEY

Computers did not force us into a new realm of the digital. Just because a particular phenomenon, whether it's a CD or an email, is made of *bits and bytes* or 1s and 0s doesn't mark the digital as a realm entirely different from the pre-digital. The digital was hidden in the analog world long before computers rolled around—in the formal logic of Smullyan's truthteller puzzles, for example. The impact of computation lies not in 1s and 0s, but in a related but very different binary: true and false.

For a computer programmer, the two binaries of true/false and 1/0 quickly become interchangeable. True is 1 and false is 0.* Most com-

* *Very* occasionally it is the other way round, to the endless vexation of programmers. I experienced that kind of vexation when programming in the ML language, which demands that negative numbers be written not with a dash but with a tilde: "-2" must be written as "~2," or

puter languages make an explicit distinction between a Boolean variable (which can be true or false) and a numerical variable (which can be 1, 0, or some other number), but in practice this division blurs. This owes to the incredibly tight relationship between logic and mathematics in computation.

To take a step back: Boolean logic is comprised of a series of logical operators, such as AND, OR, and NOT, that are applied to one or two binary variables. The expression (p AND q) is true only if p and q are both true; (p OR q) is true if p or q or both are true; (NOT p) is true if p is false and vice versa. "My name is David AND I am ten feet tall" is false. "My name is David OR I am ten feet tall" is true.* Smullyan's puzzles are disguised Boolean logic problems, the goal being to determine which p's and q's are true and which are false.

Boolean operations form the fundamental structure of computer hardware itself. Transistors on a computer chip form *logic gates,* which instantiate Boolean operators in physical hardware. Like the operators, logic gates take inputs that are each one of two possible "signals," and output one of the same two possible signals depending on those inputs. A NOT gate on a transistor will output a high voltage signal when no such signal is coming in from its input, and vice versa.† An AND gate will output a signal when current is flowing into both of its two inputs. By chaining these gates together, simple mathematics is possible. Addition and subtraction in turn permit multiplication and division, and from there on, things look less like logic and more like programming (including software implementation of Boolean logic).

There's nothing "true" or "false" on those circuit boards. The binary is between the voltage being ON and OFF, not TRUE or FALSE or even 0 and 1. No one is actually interested in whether gates output a voltage signal *except inasmuch as that signal represents something else.* And what such signals represent in a computer is an abstract notion of logi-

---

else the language will not understand you. There are reasons for this syntax; the question is whether *any* such reasons could justify violating the ubiquitous convention.

* It would be true even if I were ten feet tall, since OR means that one or both statements are true. The XOR, or exclusive-OR, operator specifies that *only* one of the two statements is true, while the other must be false.

† Generally, the possibilities are either a high voltage or a near-zero voltage, but what's important is that there are exactly *two* distinct signals—a binary representation.

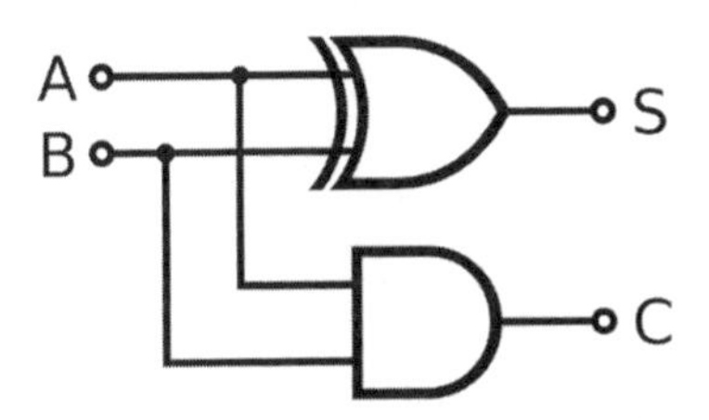

A binary half-adder, composed of an XOR gate on top and an AND gate on the bottom, that "adds" binary inputs A and B together, producing a binary sum (S) and a carry (C) output

cal truth and falsehood, a notion imposed by us. Truth and falsehood are conceptions that we apply to the presence or absence of a signal at a particular point on a circuit board.

This binary representation of truth is so ubiquitous that we chronically conflate logical (true/false) and computational (1/0, or on/off) binaries. A programming language permits me to have a Boolean variable—call it IsThisThingOn—which can be either TRUE or FALSE. The simplest representation would be to use 0 for FALSE and 1 for TRUE. But what if, with this numerical representation, we run across a 2, or some other number that's not 0 or 1? We could simply declare these other numbers to be meaningless, neither true nor false, but we do not want a trinary logic in which there are not two but three options, true, false, and meaningless. In C, the convention is that 0 is FALSE and *any other number* is TRUE. This representation appeals to our common convention of equating truth with existence. Falsehood is nothing, absence, emptiness: 0. Truth is presence, validity, existence: anything other than 0, even a negative number.

A programmer is conditioned to treat TRUE and FALSE as synonyms for 1 and 0 respectively. TRUE and FALSE become the two possible values for a single bit in a two-value system. But those TRUEs and FALSEs remain a purely formal convention, even though the words contain metaphysical weight within them. If I write `GodExists = TRUE`, that obviously doesn't cause God to exist. Nor does it cause the computer to think or believe that God exists. All it does is set a signal somewhere deep in the recesses of a computer's memory chips. That line of code, however, suggests more than just the changing of signal on a circuit board. Any meaning we assign to TRUE and FALSE besides a signal's being on or off is interpretation. When I say "God exists," I am making a claim about the real world. When a program contains the statement `GodExists = 1`, it merely indicates the alteration of a signal.

Computers do not understand the difference between TRUE and FALSE, nor do they understand what these two concepts mean. If a computer is to gain any grasp on truth as we understand it, it will have to be someplace other than in its bits.

True and False

> Who claims Truth, Truth abandons.
>
> —THOMAS PYNCHON, *Mason & Dixon*

The history of logic has long wrestled with the problem of getting that TRUE to mean more than 1 or ON. In other words, the problem of getting TRUE to say something *about the world.* The difficulty started long before computers. We may think that we can patently tell the difference between a true statement and a false statement, but outside of logic puzzles, pinning down this distinction proves horrendously vexing, to the point that both Western philosophers like Sextus Empiricus and Eastern philosophers like Nagarjuna questioned whether the truth of *anything* could be known. These concerns, which might seem like philosophical navel-gazing, became far more urgent once formal logic entered the picture. The great logician Alfred Tarski, a great influence on computer science and artificial intelligence, coined "Convention T" as a minimal criterion for the truth of a statement. Here are two examples of Convention T:

1. "The sky is blue" is true if and only if the sky is blue.
2. "I am a jelly doughnut" is true if and only if I am a jelly doughnut.

A great deal of philosophical ink has been spilled on Convention T. But I think most lay readers will agree that this criterion is not especially helpful as a practical guide to determining the truth of a statement. And we may even despair at how little progress Convention T appears to make over Aristotle's definition (itself based on Plato's) of over two thousand years ago:

> To say of what is that it is not, or of what is not that it is, is false, while to say of what is that it is, and of what is not that it is not, is true.

Yet whatever truth may be, we can safely conclude that in order to interact well with the world, computers must be able to distinguish between true and false in the same way as humans do. The inadequacies of logicians' attempts to collapse the difference between a logical truth and a worldly truth have been made all too clear by the computing age. Mathematician and phenomenologist Gian-Carlo Rota put this bluntly:

> Mathematicians are therefore mystified by the spectacle of philosophers pretending to re-inject philosophical sense into the language of mathematical logic. . . . The fake philosophical terminology of mathematical logic has misled philosophers into believing that mathematical logic deals with the truth in the philosophical sense. But this is a mistake. Mathematical logic deals not with the truth but only with the game of truth. The snobbish symbol-dropping found nowadays in philosophical papers raises eyebrows among mathematicians, like someone paying his grocery bill with Monopoly money.

Rota here echoes Plato, who was one of the first to find that truth is a *practical* matter, a matter of action rather than of pure theoretical abstraction. In Paul Friedlander's summary, "Truth, in Plato's system, is always both: reality of being and correctness of apprehension and assertion."* The troublesome gap is not between logic and language, but between logic and reality. Symbols and proofs cannot close that gap on their own.

In formal logic, truth is a construct devoid of external meaning. By claiming that "X = X" is true for any possible value of X, I am applying

* Or, in William James's words, "Truth *happens* to an idea. It *becomes* true, is *made* true by events."

the axiom of identity to the logical system: it is always true that something is equal to itself. Even trying to apply this simple principle in reality quickly runs up against ambiguities. If I say "David = David," am I saying the same kind of thing as "2 = 2"? Or does using a person's name instead of a number change the sort of truth in play? It doesn't seem possible that 2 could not equal 2, but could David not equal David? What if there are two different Davids involved? Context comes crashing in very quickly.

The pioneering *philosophical* logicians—Gottlob Frege, Bertrand Russell, Ludwig Wittgenstein, Rudolf Carnap—joined the formalism of logic with the content and metaphysics of reality.* For them, logic had to have some regulating function on the world. Carnap titled his magnum opus *The Logical Structure of the World,* and his ambition was to show that a universal language of logic could wholly represent the world. This was a utopian view for Carnap, who was a committed socialist and pacifist. In the introduction to *The Logical Structure of the World* (1928), he made clear that the larger purpose of his work was to set the world aright and eliminate the misunderstandings and irrationality that give rise to suffering:

> We feel that there is an inner kinship between the attitude on which our philosophical work is founded and the intellectual attitude which currently manifests itself in entirely different walks of life; we feel this orientation in artistic movements, especially in architecture, and in movements which strive for meaningful forms of personal and collective life, of education, and of external organisations in general. We feel all around us the same basic orientation, the same style of thinking and doing. It is an orientation which demands clarity everywhere, but which realizes that the fabric of life can never quite be comprehended. It makes us pay careful attention to detail and at the same time recognizes the great lines which run through the whole. It is an orientation which acknowledges the bonds which tie people together, but at the same time strives for free development of the individual. Our work is carried by the faith that this attitude will win the future.

* Arguably nineteenth-century American philosopher Charles Sanders Peirce belongs on this list too, but his work did not find influence for many decades.

Carnap's attitude remains close to the hearts of many engineers, scientists, and mathematicians. It is the vision and hope that the world can one day make as much sense as logic and mathematics do. The promise of mathematics, as held out by Plato so long ago, is a vision of a harmonious world where misery, misunderstanding, and conflict are logical contradictions to be ironed out through the use of reason.

Wittgenstein's opening line of the *Tractatus Logico-Philosophicus* reads, "The world is all that is the case." It's a claim that also joins logic and the world. "There's a unicorn in my room" is false because there is no unicorn in my room. But what about a statement like "All men are created equal"? How can one establish the truth of something like that? Computers cannot yet even detect whether there is a unicorn in my room.

We know what truth and falsehood are. Or, at least, we think we do. When it comes to anything beyond the simplest facts, we have a much harder time settling on truth in our daily lives than the average scientist does at getting to an accepted truth in his or her field.* It takes an entire legal system to determine who caused a car accident. My wife says it's a chilly day while I find it temperate. We use words like "old," "thin," "ugly," and "dumb" haphazardly and inconsistently, and we live with disagreement over when they should be applied. In everyday life, the line between fact and opinion is so blurred as to be frequently invisible, and our speech is beset by chronic vagueness and ambiguity, which generates as much confusion as it does richness. Code flounders on ambiguity, while literature thrives on it.

* Robert Musil remarks on science's remarkably prolific and satisfying track record of generating "truths": "If we translate 'scientific outlook' into 'view of life,' 'hypothesis' into 'attempt,' and 'truth' into 'action,' then there would be no notable scientist or mathematician whose life's work, in courage and revolutionary impact, did not far outmatch the greatest deeds in history. The man has not yet been born who could say to his followers: 'You may steal, kill, fornicate—our teaching is so strong that it will transform the cesspool of your sins into clear, sparkling mountain streams.' But in science it happens every few years that something till then held to be in error suddenly revolutionizes the field, or that some dim and disdained idea becomes the ruler of a new realm of thought. Such events are not merely upheavals but lead us upward like a Jacob's ladder. The life of science is as strong and carefree and glorious as a fairy tale." Life, whether in the words of philosophers or poets, does not generate revolutions and discoveries in the way that science does. Any popular such "discovery," like Nietzsche's Death of God or Warhol's Fifteen Minutes of Fame, becomes subject to debate and analogy. Science has many battles, but the expectation is always that any scientific idea is bound to either be true or false, and that we will know which sooner or later. It is an unfortunate but understandable side effect that many scientists believe such definite answers can also be obtained on political and moral questions.

Even in science it's tricky enough. Newton's laws were accepted as truth for hundreds of years. They are still useful, but we now know they are not universally true. For high masses and high velocities, Newton's laws fail. Newton didn't imagine that the speed of light would function as any sort of absolute speed limit. He didn't know what light was, much less that it moved at any speed. Newton's laws are true sometimes, with caveats. But is a truth with caveats still true?

All worldly truths have contexts, conditions, and caveats. Wittgenstein suggested that worldly contexts, conditions, and caveats were precisely those of formal logic: "The limits of logic are the limits of my world." Wittgenstein looked to logic to find how truth works in the real world. Our linguistic statements, he proposed, depict true or false states of affairs, and formal logic provides the rules that regulate our construction of these statements. Language and the world share logical form, which is also the form of reality. This attempt to regiment language as formal logic went on to be an article of faith for many computer scientists and cognitive scientists for decades, as well as exerting a foundational influence on Noam Chomsky's linguistics.

But after a ten-year break from philosophy, Wittgenstein changed his mind. Language did *not* have such a fixed, logic-bound relation to reality. The process of "measuring" the truth of a statement against reality was neither objective nor cleanly delineated. The meaning of what we say can't be abstracted away from the context in which we say it: "We are unable clearly to circumscribe the concepts we use; not because we don't know their real definition, but because there is no real 'definition' to them." Instead, what we say and the *truth* of what we say are grounded in a set of social practices. In speaking, there clearly are rules that shouldn't be broken and clearly ways of speaking that are blatantly incorrect, though they can change over time and admit to flexible interpretations even on a daily basis. It's just that explicitly delineating those boundaries is extremely difficult, because language is not built up through organized, hierarchical rules, but through byzantine, overlapping practices. Some truths *can* be pinned down with practical certainty, just *not in isolation and not without context*. Truth is not necessarily relative, but neither does it exist in a vacuum.

Formal logic does. The context of formal logic is austere and explicit, which is what makes it amenable to computers. But the cost of that simplicity is giving up the link between logic and reality. The true and false of formal logic are not the true and false of which we speak every day. The true and false of formal logic are empty abstractions: 1s and 0s.

I believe in an objective external reality to which we all must conform, but I am also a social constructionist. We *negotiate* truths as much as we discover them. We invent and create a language to describe that reality, but as Wittgenstein found, that language has a somewhat haphazard relationship to reality. So what we call "truth" is never quite *the* truth, but a representational approximation of it.* We generate and agree upon everyday truths, and we fight over them. At the end of the day, if we're lucky, consensus prevails. What we call "truth" remains loosely in line with reality because practical usages weed out falsehoods. Astronomy proved more useful than astrology, so astrology diminished and astronomy grew. Aristotle's physics did not work for building machines and predicting movement; Newton's did. Humanity had to *agree* that Newton's physics were true and Aristotle's were false, but we had good reason to do so. Computers do not have *reasons* for what they deem true or false. To know the truth of something is to be able to explain it, use it, argue for it, and revise it in light of new evidence. When computers can use what they *believe* to be TRUE to function (or malfunction) in the world, and modify and judge these "truths" based on their interactions with the world, that is when we'll know that computers have an understanding of what we mean by truths and falsehoods. That is when they will start to become human.

* In William C. Wimsatt's wonderful description, scientific theories are "piecewise approximations to reality."

INTERLUDE:

FOREIGN TONGUES

> Exactly I and exactly the world
> Fail to meet by a moment, and a word.
>
> —LAURA RIDING, "The World and I"

COMPUTERS LACK SKILL with the interface—language—that humans use to tame and conceptualize the world. I sympathize with computers in that regard. I was four when I first spoke in English. I was six when I first programmed in Logo. Brain development varies wildly among children. Math and science came relatively easily to me; human language has always been harder. Foreign languages come slower to me than to most. English feels no more like a native language to me than Logo and C. I am ill at ease in a room of people speaking quickly and fluidly.

Perhaps as a consequence, I have kept my feet in multiple social environs simultaneously, most often through a combination of humanities and technology work. I read Ralph Ellison while learning C, Robert Musil while writing a compiler, James Joyce while working at Google. The communities were complementary. At Google we could have a long discussion of New York subway optimization problems—where to stand on the platform for the emptiest car, when to give up on waiting for a train. In academia, abstract and abstruse discussion of the sociopolitical implications of Hegel's master/slave dialectic were mumbo jumbo to tech wonks. This theorizing constituted a style of loosely analogical thinking that didn't mesh with the precise engineering of my workday.

These two groups by and large thought ill of each other. My caricature, while exaggerated, is not too far off: to tech wonks, humanities scholars build ill-founded castles in the air with meaningless words to prove that nothing means anything.* To the humanities scholars, tech wonks are imprisoned by a positivistic mind-set that leaves little capacity for context, speculation, or modes of thought that cannot be reduced to logical form. Each side tends to be remarkably uncharitable to the other. The playing ground is hardly level, however; the tech boom and increasing centrality of computation to life gave the tech culture a sense of relevance and financial success that contrasted to the glut of low-paid labor in academia. Academics can no longer ignore tech, and fads like the "digital humanities" as well as technology studies became new mechanisms of propping up the precarious system of American higher education.

To my mind, the two domains were equal—and equally foreign. The exactitude of computer science provided me with useful checks on linguistic hot air. Humanistic fancy, however, enabled me to figure out what I was doing in this technocratic labyrinth, and to ask myself why I was doing it and where it was going. I no longer program full-time, but I miss the mental practice it gave me, which served to focus my mind in a rigorously geometric fashion.†

I didn't belong to either community. I was a poor fit whether the topic of conversation was econometrics or Hegel. When I think of my social life in childhood, I remember silence, punctuated by occasional, apprehensive experimentation and subsequent failure. I dub it failure not because I failed to fit in, which is every child's experience, but because none of my experiments seemed to result sufficiently in my being myself. My words were never quite right, and people never seemed to

* Alan Sokal's *Social Text* hoax remains the perfect illustration. While I find much of worth in the larger fields of philosophy and literature, Sokal picked very ripe targets that had, for the most part, become exactly the self-parodies that critics thought them to be. That some academics still get defensive when it's brought up indicates just how close to the bone he hit.

† To exercise that part of my brain I now turn to half-understood books on quantum physics, machine learning papers, and Stephen Lavelle's diabolical puzzle game, *Stephen's Sausage Roll.*

understand what I said in the way I had intended. I felt poorly served by the English language, and well into my teens I saw few models for how I could bend and mold it to the shape of my mental images and feelings. My favorite authors became those, like Virginia Woolf and Ralph Ellison, who were practiced in ventriloquizing through a wide variety of voices and putting each one into question through its proximity to others. But before I had those reference points, I found myself silent. Most of all, I was *watching:* I was Camus's judge-penitent in *The Fall.* I was the watcher in Kafka's "At Night":

> Deeply lost in the night. Just as one sometimes lowers one's head to reflect, thus to be utterly lost in the night. All around people are asleep. It's just play acting, an innocent self-deception, that they sleep in houses, in safe beds, under a safe roof, stretched out or curled up on mattresses, in sheets, under blankets; in reality they have flocked together as they had once upon a time and again later in a deserted region, a camp in the open, a countless number of men, an army, a people, under a cold sky on cold earth, collapsed where once they had stood, forehead pressed on the arm, face to the ground, breathing quietly. And you are watching, are one of the watchmen, you find the next one by brandishing a burning stick from the brushwood pile beside you. Why are you watching? Someone must watch, it is said. Someone must be there.

I marveled at practiced speakers who could give the same stump speech on whatever topic of their choosing (the mind-body problem, the Republican majority, the Turing Test) without so much as thinking about the words they were saying. Words never stuck in my mind—only ideas, and ideas can't be spoken without words.

Heinrich von Kleist suggested that ideas of the mind were alive, which made them unable to be captured with a rote set of words. Ideas that admitted to a single permanent expression were dead—in the air and on the page.

> Speech then is not at all an impediment; it is not, as one might say, a brake on the mind but rather a second wheel running along parallel on the same axle. . . . For it is not *we* who know things but pre-eminently a certain *condition* of ours which knows. Only very

> commonplace intellects, people who yesterday learned by heart what the state is and today have forgotten it again, will have their answers pat in an examination.

As I grew more fluent in English, I still found it hard to keep up with the parade of styles required by social norms. In college, I found that the freewheeling mode of essay writing I preferred did not make a good impression on most professors. There were exceptions, which made it that much more baffling. Most did not welcome the confused, inchoate flood of ideas that my teenage self liked to set down, and I was asked to adopt a very precise and academic subdialect to get high marks. This dialect varied by discipline and even by class. Many of my humanities classes were exercises in deriving the dialect required and forcing one or two ideas into the necessary framework, throwing away 90 percent of what I had found interesting. But faced with the choice between that and the seemingly impossible task of creating work that was simultaneously satisfying both to me and to the professor, I thought a good grade would serve me better in life. I could write eccentrically in my spare time. By the time I needed to decide between a job in programming or going to graduate school for philosophy or literature, the choice was obvious.

Computer languages are frequently variations on a theme. There are a few paradigms, such as the procedural, imperative structure of C (and its progenitor Algol) or the functional structure of LISP, which subsequent languages have refined and fine-tuned to bring out certain strengths at the expense of others. Imperative languages are structured as a sequence of commands that affect their surrounding execution environment; functional languages are structured as the evaluation of algebraic functions. The paradigms match different tasks better or worse than one another. So it is with human languages. Certain modes of speech and thought work better than others in certain contexts, so I learned how to speak in different settings. Algorithms can be expressed in different programming languages. A sorting algorithm is the same whether in C or LISP, even if its code is wholly different. I did not wish to give myself over to one particular dialect so that my use of language became predictable and ossified. I balanced the tech world with graduate school and tech articles with fiction.

I found myself enduring the constant push-pull of withdrawing

from a social context while respecting its norms and its participants.*
My wife has observed that in raw form, my writing often reads like it was translated from German—all complex noun phrases and dependent clauses. I am aware that I am putting forth a great deal of effort to communicate simply and clearly. The autistic hacker Meredith L. Patterson explained her social difficulties in this way:

> While LiveJournal and Twitter have taught me to translate my thoughts into English faster and faster over time, translation is still *hard* and it still isn't realtime . . . those childhood lessons about not being so rudely direct have stuck.

Unlike Patterson, I lacked *any* dominant instinct. I had different reference points for this sort of social autodidacticism, one being Denis Diderot's *Rameau's Nephew,* in which a bon vivant discusses how he has learned the arts of conversation, flirtation, and social parasitism through hard work and study:†

> I find in writers a digest of everything one ought to do, and everything one ought not to say. . . . I am myself, and that is what I shall remain; but I behave and talk in a socially acceptable manner.
>
> And by the way, you shouldn't suppose that I'm the only reader of this kind. The sole merit I claim here is having accomplished systematically, through clear thinking and rational, accurate observation, what the majority of others do by instinct. That's why their reading doesn't make them better than me, but instead, they go on being ridiculous, whereas I am so only when I mean to be, and then I leave them far behind me; for the same art that shows me how to avoid ridicule in certain situations, shows me also, in other situations, how to achieve it at a superior level.
>
> Then I bring to mind everything others have said, everything I've read, and I add everything of my own invention, which in this domain is surprisingly abundant.

* Or, as the cartoonist Eddie Campbell puts it, the tug between the Scylla of compromise and the Charybdis of failure.

† A writer's writer, Diderot had a heavy influence on Goethe and Hegel, who cited *Rameau's Nephew* in *The Phenomenology of Spirit,* which in turn influenced Lionel Trilling's *Sincerity and Authenticity.*

I used to look down on this sort of apparent fakery, prizing some ideal of honest and earnest self-expression. I slowly came to see, though, that it was not fakery but communication that I was after, and that only by learning such mechanisms could I be understood. To put it another way, I gave up feeling entitled to be understood on my own terms. And so I adopted stylistic dialects just as a computer runs different applications and compiles different languages into machine code.

The character of Rameau's nephew is a general-purpose machine that runs varieties of social interaction software. I have strived to be similarly adaptable. But to be a universal social machine is also not to have the immediacy of instinct and the rushed emotions of saying what I mean and meaning what I say. That translation puts a kink in the process. I am slower and more error-prone than machines that run a particular social language natively.* The writer's eternal demon—do these words mean what I think or are there better words for it?—was with me long before I became a writer.

The hesitation I feel in using particular words or idioms, and the uncertainty I have that they ever truly fulfill my intentions, are not merely private matters for me. Even as I negotiate between my words and my thoughts, there is a much larger, but parallel task facing society: the primitive, dissonant relationship between *our* words and our computers' code. We have been teaching computers how to translate crudely between human language and computer code. What they are able to understand and misunderstand now has the power to shape our lives.

* Ingeborg Bachmann's short story "Simultan" (translated as "Word for Word"), a portrait of the inside of a translator's mind, is as good a depiction of this non-native mind-set as I know.

PART II

4

NAMING OF PARTS

Labels

> Words are not a medium in which to copy life. Their true work is to restore life itself to order.
>
> —I. A. RICHARDS, *The Philosophy of Rhetoric*

MY NAME IS DAVID. It derives from the Hebrew word דוד, which is commonly translated as "beloved," sometimes as "uncle." My parents chose it because it was popular at the time, it was my father's middle name, and it sounded good to them. To most people my name means nothing. It is merely what I am called. But I could also be called a human, a programmer, a writer, a husband, a father, a New Yorker, a Jew. These words have meaning, and do not *identify* me as much as they *describe* me. They are *labels.*

For much of history, people and things have acquired labels haphazardly and often arbitrarily. Computers change that. Computers do not interact with people and things. They operate only on their internal *representations* of people and things. Sometimes the representations are mathematical or schematic, as with a CT scan or a weather model, but often the representation is merely a simple data structure named after the person or object to which it corresponds. A product on Amazon, a Facebook profile, or a music file: all are specified by the various identifiers and classifiers attached to them—these are often called metadata. The names and IDs (like a book's ISBN or a numeric user ID) may uniquely pinpoint an entity, but they don't describe it. Instead, we assign descriptive *labels* to entities. Labels are words, but they are words that do not name a particular person or thing. They are used across a wide variety of contexts. They are generic and universal placeholders.

In Genesis, God tasked Adam with naming the animals. Other animals possess the ability to correlate two stimuli together (a sound and a sight, an emotion and a situation), but only humans assign arbitrary verbal labels to stimuli. In this regard, computers are closer to animals than to humans. They associate labels with phenomena only after being trained to do so. Advances in machine learning have allowed computers to make impressive, though tentative, steps in the direction of generalizing labels to new situations, but computers are still less adept at doing so than toddlers.

Programming languages and human languages are two different mechanisms, intended for different purposes. Computer language is closer to *logic* or *algebra,* which is to say, computer languages deal ultimately with *numbers.** Human languages deal with *human life* and everything in it. Translating between human and machine languages is far trickier than translating from English into French. Whatever is said in either English or French, its speakers are still talking about the same underlying reality, as it is perceived by humans. There is a shared context, a shared knowledge of the same *world.* Computers do not share this context. To *understand* us, computers require that we build context *on top of* the raw numerical material, just as the human mind is built on top of neural cells.

First, we construct a relationship between the computer's numbers and the real world as we know and experience it. If we map numbers 1 through 26 to the letters of the alphabet and add a few more numbers for punctuation, we can easily represent the entire syntax of the English language on a computer. But if I want to communicate the message "DAVID LIKES CHOCOLATE," a long string of numbers beginning

* Or, more precisely, mathematical abstractions, whether numbers or bits or sets or functions or spatial figures or what have you. These all fall under the rubric of data, but "data" has become a hopelessly vague term that goes far beyond the reach of computers, so I will tend to say "numbers," the underlying representation for (most) computer data.

with 5-1-22-9-5 is not the most efficient or logical way to express that sentiment. Rather, I can create a custom encoding and say that the numbers represent the words as follows:

```
1 = DAVID
2 = LIKES
3 = CHOCOLATE
```

As long as a computer knows how and when to use this encoding, 1-2-3 holds the same meaning as "DAVID LIKES CHOCOLATE" with a fraction of the space required. But no computer would ever use this encoding, because it has extremely limited and specific utility. It does not generalize. Perhaps I want to talk about Daisy and her hatred of chocolate as well. I add:

```
4 = DAISY
5 = HATES
```

I then can write the sequence 4-5-3. We can go on like this ad hoc. The computer knows nothing about me, Daisy, or chocolate, but it does have a *code* (or an *encoding*) by which something about me or Daisy can be communicated to another person able to decode the message. We can build up a numerical encoding for storing and transmitting complex information. We can also enumerate categories:

```
enum TRAFFICLIGHT_STATE {
  RED    = 1,
  YELLOW = 2,
  GREEN  = 3
}
```

The computer has no conception of what it's doing when a particular spot in its memory shifts from one (**RED**) to three (**GREEN**), but if that computer is hooked up to a traffic light controller, such a change would be meaningful *to us* as designating the change of a traffic light from red to green.

That gap between seeing data *as numbers* and seeing data *for what it represents* is the fundamental difference between computer language and human language. It is the difference between a rote encoding and a meaningful conversational tool. As computers increasingly facilitate and even dominate the socioeconomic fabric of our world, closing the gap between those two kinds of languages has become an increasingly urgent concern of computer science and software engineering. If computers could bridge the gap and understand the world the way we do, we could off-load our own mental labor onto it, and perhaps leave computers to do a better job of managing large-scale economic and organizational problems. But it is easy to underestimate that gap if you are a techno-optimist—just as easy as it is to think it is unbridgeable if you are a techno-pessimist.

What's the secret to creating a *language* in which computers *do* understand what their data represents? In a nutshell, this is the problem of artificial intelligence.

We ourselves are discovering that we do not quite understand how *we* talk about things. Our brains are hard-wired to learn human language. We are born with an innate capacity for it, which other animals lack. Humans have also learned how to speak the codes of machines, that is, mathematics and logic. As we try to drag computers into the realm of human language and clumsily try to teach them a capacity for understanding, they are teaching us to think in their codes.

Much of the nuts and bolts of data processing today is done not in the arcane and ingenious algorithms that analyze data, but in the labeling required for those algorithms to be useful to us. The process of labeling is more accurately a process of *encoding;* assigning discrete identifiers to the hazy concepts that permit us to function in the world. We do not yet know how to program a computer to understand our conception of a *thing* like "dog." A concept like "dog" is a loose-knit collection of facts and images that varies from person to person. All the computer knows, to begin with, is the word itself. We *encode* this concept in three letters: "DOG." A computer would represent it as three numbers for the ASCII equivalent of those letters—68 79 71—or perhaps it would encode it in some other numerical scheme in a specific program: say a pet-classifying program, where dogs are 1, cats are 2, turtles are 3, and so on. Regardless, the computer has no conception of *a dog,* just the *encoding* for the *label* assigned to the concept. We need to *encode* the world before computers can process it.

The word "code" can mean either a series of programming instructions or the labels that an encoding contains. To keep things clear, I will use the word "label" to refer to the elements of an encoding rather than "code" or "encoding."

Computers have us thinking so much in labels today that we must remember we were the ones who assigned those labels, rather arbitrarily, in the first place. We take them for granted, but they are merely a matter of agreed-upon—or disputed—convention. Today, the origins of a complex encoding such as Western musical notation are obscured, the notation taken for granted. Even the Western alphabet, which we take to be fixed, arrived at its current stable form only because too many people were using it for local changes to remain feasible. The histories of most encodings are so gnarled that it is difficult to see past their complexities to understand how labels shape life. So consider two of the oldest and most universal labels of all.

Male and Female

> The words come rolling up to us, we must be careful not to get run over.
>
> —ALFRED DÖBLIN, *Berlin Alexanderplatz*

Some questions have more definite answers than others. There are questions whose answers lie in some underlying fact of the matter, like "Does the Earth go around the Sun or vice versa?" There are questions like "How many planets are there in our solar system?" whose answers change as criteria and terminology evolve. And there are questions where it's not clear which facts are relevant, like "When did the Roman Empire fall?" (which yields answers ranging from "476 CE" to "It never did") and "Is it proper English grammar to wantonly split infinitives?" (where some will say "Of course" and others will say "It sounds awful and déclassé"). We *make* answers to these last kinds of questions, arguing over them until there is either an authoritative consensus or people agree to disagree.

Computers, generally, do not distinguish between these types of questions. Their data rarely contains epistemological caveats. They present it back to us exactly how it was stored, which is usually with stark certainty. We tend not to turn to computers for those last sort of questions, but the ubiquity of computers forces us to answer some of those questions before they can process our data. Here is one such question: "How many genders are there?"

Western societies increasingly recognize divisions between gender and biological sex. That is, one's gender may not have any *necessary* relation to one's physiology or biology, even though it frequently correlates. But what *are* the criteria for gender? There is no consensus.* The debate around this question has split the term "gender" into many subcategories: gender identity, gender role, gender presentation, gender expression. Yet we are so incessantly identified by our sex and/or gender that we are all forced to pick *one* label under which we will be classified. Most people are assigned the label of male or female at birth and stick with it. Some do not. Outside the two dominant categories, today there are various flavors of transgenders, genderqueers, agenders, androgynes, pomosexuals, and others—to say nothing of intersex people, who despite being classified by physiology are often spoken of as being their own gender. Science does not provide definitive answers here. If, as the World Health Organization states, gender is a social construct that varies by culture and era, there is no scientific fact as to how many genders there are. There are only people's opinions as to how many there *ought* to be.

Computers can easily handle the addition of new categories, simply by extending the encodings of sex or gender to encompass whatever labels we think they should contain. When Facebook changed their list of gender choices from two to fifty-one in 2014, then subsequently added seven more (twenty more in the UK), Facebook wasn't trying to prescribe a new taxonomy of gender. But it did just that. With approximately a billion users for whom they are setting the rules, Facebook can't avoid being prescriptive. That's why activist groups lobbied Facebook to, first, add additional genders and then to let people type in their own. Google, whose social network does not hold much sway in

* The debate around transmedicalism is only one example of how unresolved these questions are.

our lives, merely offers "Male," "Female," and "Other," but Facebook at the time of this writing offers:

Agender
Androgyne
Androgynous
Bigender
Cis
Cis Female
Cis Male
Cis Man
Cis Woman
Cisgender
Cisgender Female
Cisgender Male
Cisgender Man
Cisgender Woman
Female
Female to Male
FTM
Gender Fluid
Gender Nonconforming
Gender Questioning
Gender Variant
Genderqueer
Intersex
Male to Female
Male
MTF
Neither
Neutrois
Non-binary
Other
Pangender
Trans
Trans*
Transsexual
Transsexual Female

Transsexual Male
Transsexual Man
Transsexual Person
Transsexual Woman
Trans Female
Trans* Female
Trans Male
Trans* Male
Trans Man
Trans* Man
Trans Person
Trans* Person
Trans Woman
Trans* Woman
Transfeminine
Transgender Female
Transgender Male
Transgender Man
Transgender Person
Transgender Woman
Transmasculine
Two-spirit

Facebook originally forced users to select exclusively from this list, but subsequently let users type in whatever terms they want, though the interface strongly discourages doing so. (It took me a few minutes to figure out how to do it.) Users may choose up to ten terms to describe themselves, with no restrictions on contradictions. I've set mine to "Male, Female, Neither, and None." By offering a strictly defined set of two, fifty-one, fifty-seven, seventy-one, or any other number of genders and mandating that every user choose one, Facebook ensures that its data analysis will be neat. Without a default set, Facebook would be left with a long tail of rare, sometimes unique, and unstandardized genders. The data would be messier and harder to analyze.

Facebook does not, however, encourage going outside the dominant two categories. Users have to choose male or female when they sign up for an account, then they must change their gender on their profile pages. And Facebook asks users to choose a preferred pronoun, for which there are only three options: male, female, and "neutral" ("they"). As we'll see later, for all the praise *and* criticism that Facebook's expansion has received from activists on both sides of the fence, Facebook's ultimate commitment to these new labels is rather superficial indeed.

Facebook tried to accommodate an increasingly flexible gender taxonomy. Computers are not terribly good with flexibility. Yet there is a feedback mechanism at work here. The demand to *explicitly* categorize one's self within a particular typology, whether to Facebook or to corporate management, is apt to foster dissatisfaction with that typology. Gender is a construct that is continually reified—but also continually criticized.

The danger of reinforcing one specific typology is that typologies come to look antiquated and prejudiced in the long run. Our racial categories may very well evolve from what they are right now, and a lot of current thinking about race may be considered outdated and ignorant in a few years' time. "Conservative" and "Liberal," labels that Facebook secretly assigns to its users, have very different meanings today than they did twenty-five years ago; their definitions will undoubtedly continue to evolve. In such a dynamic world, computers paradoxically enable us to revise and refine our categorizations even as they insist that we continue to *make* those classifications. To be described to a computer is to be described by labels. To be described by labels is to make a *selection* from categories. The implicit is not so much disallowed as it is overshadowed. Data prefers to be explicit.

Some of these typologies are obscured from users, who are quietly classified without their knowledge. But like gender, many of these labels are unstable. Facebook deems me a member of Generation X, which it defines as people having been born between 1961 and 1981. A quick glance at the literature shows the end date as early as 1964 and as late as 1984. Growing up, I was told I was too young to be Generation X, but at some point, the collective hivebrain of marketers decided differently. I was not Generation Y (later redubbed millennials); I was Generation X after all. If they change their minds again, will Facebook catch up? More likely, Facebook itself will help select the particular date range by virtue of their sheer market power. It's not the truth; it's just the force of Facebook's version of the truth. Facebook is now trying to determine the difference between "fake news" and "real news," and between "political information" and "political propaganda." Those distinctions, too, are Facebook's versions of the truth, defined not only by Facebook but by third parties trying to shape and influence users through Facebook advertising and applications.

Moreover, discrete categories ignore variation within those categories. Here's an example. I am one of the proud 10 percent of left-handed people, as are both my siblings and one of my children. Both of my parents are right-handed, though my father, like many of his generation, was forced to write (rather poorly) with his right hand in school, so we believe that he is latently left-handed. I associate being left-handed with smearing ink and graphite on my pinky finger and with uncomfortable right-handed desk-chairs.* Let us say that Facebook requires us to specify our handedness, and let us say that this is something I care about. I would certainly select left-handed, but maybe I would then want to elaborate and explain that I do not think the dominance is strong. My right arm and hand are stronger than my left. Whereas my left hand suggests dominance by my right brain, my right eye is my dominant eye, governed by my left brain. In first grade, frustrated with the terrible single pair of left-handed scissors in the classroom, I forced myself to use the crisp, new right-handed scissors they had, and I never went back. I use a mouse with my right hand but tend to use the phone with my left. None of this changes my self-classification and societal classification as left-handed. But if society were, for whatever reason, to start *caring about* and *refining* those labels, as it has with gender and race, those nuances would be impossible to capture on Facebook.

Cultural change drives the revision of labels. A stagnant culture allows for static (and stagnant) labels. A fluid culture permits the current classification to be torn up and replaced by the next, equally contingent classification. Stasis gives the illusion of permanence. And the more a society reinforces particular taxonomies, the more inertia these taxonomies create against social change.

As a consequence of mass media and global interconnectedness, today's societies are far more dynamic than their predecessors. Sub-

* The studies claiming that left-handed people die earlier than right-handed people are mostly bunk. One study showed that they died younger only because they couldn't find *any* left-handed people above a certain age. This, it turned out, was because they had all been forced to be right-handed in school during the relevant time period.

cultures meet, exchange ideas, and evolve at a rate that was unknown before the invention of the railroad, and that was subsequently amplified by the automobile and airplane. The internet is only the latest upshift to the speed of cultural exchange. We are constantly inundated with new cultural material that obscures the fact that a great deal of humanity's social heritage emerged from vastly more static societies. These older heritages will not adapt to our current high-velocity culture without severe modifications.

People may not be using terms like "pangender" or "biracial" in fifty years, much less two thousand. But a label like "Confucian," which calmly escorted the regimented bureaucracy of the Chinese empire, managed to retain a complex conceptual meaning for centuries.* Is it possible for anything to have that sort of longevity today? Since computers demand that we label ourselves, our beliefs, and everything in the world around us, how will technology impact the labels that we use today, and how durable will those labels prove to be?

Masterminds and Crackpots

> In some sciences, the endeavor to discover a universal principle may often be just as fruitless as the endeavor of a mineralogist to discover some primary universal element through the compounding of which all minerals arose. Nature creates neither genera nor species, but individua, and in our shortsightedness, we must seek out similarities to be able to retain many things simultaneously. These concepts become more and more inaccurate the broader the categories are which we create.
>
> —GEORG CHRISTOPH LICHTENBERG

* Chinese civilization and culture hold a record for stability and continuity that puts any other large region of the world to shame. It's a testament to the robustness of Confucian thinking that it remains standing in somewhat recognizable form after almost 2,500 years. As a cultural entity, Christianity probably comes closest for sheer duration, but the multitude of belief systems and cultural movements called "Christianity" and the varieties of ideas and practices attributed to the name are so wildly disparate that you'd be hard-pressed to identify *any* commonality between them short of their veneration of a person whose birthdate (approximately) sets the inflection point of the world's dominant calendar.

My name is David, and I am a green. I discovered this at a corporate training retreat where the hosts asked me and my coworkers to take a test that classified each of us as one of four personality types. They were identified by color: blue, green, gold, orange. There are many variations on this sort of classification, but this test described the colors more or less like this:

GREEN: Analytical, logical, theoretical, introverted engineer
BLUE: Harmony-oriented, compassionate peacemaker
GOLD: Traditional, conventional, structure-loving bureaucrat
ORANGE: Outgoing, creative, upbeat, visionary leader

At this gathering of Google employees, greens overwhelmingly dominated, and I was one of them. There were smaller groups of blues and oranges, and maybe one or two golds. The organizers told us that every Google retreat ended up being mostly greens.

This classification is not science. It does capture *something* about the differences between people, but what it captures is arbitrarily selected and minimal. Who drew the dividing lines? The corporate consultants behind these sorts of tests claim they harken back to the ancient Greeks and Hippocrates's classifications of the four humors, or temperaments. Some versions map the humors like this:

GREEN: phlegmatic (water, phlegm)
BLUE: sanguine (air, blood)
GOLD: melancholic (earth, black bile)
ORANGE: choleric (fire, yellow bile)

While the theory of the four humors continues to inspire artistic works like Paul Hindemith's *The Four Temperaments* and Carl Nielsen's second symphony, we no longer look to Hippocrates for medical guidance. Why look to him for personality classifications? Because humans like simple typologies, and fourfold classifications are easy to grasp. Division into two parts is too simple.* Dividing things into three makes

* Traditionally, almost any folk dichotomy sooner or later comes to echo the primeval dichotomies of male and female, light and dark, and life and death. These are fairly loaded dichotomies that are best avoided when one is seeking a supposedly neutral typology.

Leonhard Thurneysser's alchemical depiction of the four humors in *Quinta Essentia* (1574)

symmetry impossible: dynamics reduce to two against one.* But four is aesthetically satisfying as well as flexible. A fourfold grouping can be carved up into pairings *and* oppositions, it can model symmetrical pairings (2 vs. 2) as well as unequal conflicts (3 vs. 1), and each of the four elements can have similarities with and differences from the other three. Hippocrates's typology stuck in part because it was memorable, and folk theories worldwide likewise embrace fourfold classifications of everything from elements (earth, air, fire, water) to causes (material, formal, efficient, final) to Indian varnas and castes to Mencius's virtues (benevolence [rén], righteousness [yì], wisdom [zhì], and propriety [lǐ]) to Buddhism's four noble truths.†

* Tripartite schemata are fascinating. They discourage opposition and symmetry, while encouraging instability and conflict. Things are frequently divided into three, whether Plato's tripartite soul or the Christian Trinity or Freud's psyche, but things are far less often classified into one of three *separate* categories.

† Anecdotally, Chinese culture seems to have a fondness for fivefold structures: Confucius's five bonds, the five elements of Wu Xing, and the five elements of qi in Chinese medicine. East Asian cultures hold a great deal of numerological superstition, particularly around the number four (which is homophonous with the word for death), so perhaps this has served as a disincentive against fourfold structures. Declaring such a low number to be unlucky is a tremendous inconvenience; far more buildings have fourth floors than thirteenth floors. Better to make large numbers unlucky: I can't think of too many occasions where superstitions

The virtues of a fourfold classification, however, are all artifacts of the human imposition of arbitrary, simplifying order onto a far more complex reality. There is no reason why our cognitive preferences should lead us to anything resembling an *accurate* representation of personality types. But we easily grasp neat fourfold classifications, and so they stick. We can't help ourselves. Humans are excellent at conjuring meaning. We inflate the most arbitrary of symbols (counting numbers, for example) with multitudinous and conflicting associations, then hold fast to them until a new system topples them. But our cognitive biases weigh heavily in the construction of these systems.

The surfeit of fourfold personality schemes circulating today mostly derives from the unscientific, speculative, yet remarkably successful efforts of two amateur psychologists named Myers and Briggs. The Myers-Briggs Type Indicator (MBTI) test has four binary axes that allow people to classify themselves into one of sixteen types (2^4). There is little to indicate that this has any scientific validity whatsoever, but the system, sometimes termed the thinking person's astrology, undeniably appeals to a great many people. At least among my generation and demographic, I've seen people on dating sites, on Twitter, on Facebook, on Quora, and in casual conversation invoke their MBTI type as a shorthand for their personality: "Oh, I have to do it my own way because I'm an INFJ"; "Sorry I was such a jerk; I can't help it, I'm an ENFP." Surely something so reductive couldn't indicate much substantive about a person, could it? So why is it that the typology is so popular? Why do people accept it, and what purpose does it serve to them? Businesses and career development organizations use MBTI and MBTI-like typologies to predict in which role a person will fare best: research, communications, support, etc. The MBTI publisher Consulting Psychologists Press, which markets and administers the personality test to corporations, universities, and governments, claims that over 80 percent of Fortune 500 companies assess their employees with the MBTI. Despite long-standing skepticism from professional psychologists, it can be hard to avoid being classified by the MBTI.

The MBTI itself has amassed surprising authority through a long,

---

around the number 666 have come into play. The prize for inconvenience must go to the Russian superstition that and even numbers of flowers are unlucky, forcing purchasers to count buds lest they give the recipient an even number.

What's Your Personality Type?

Use the questions on the outside of the chart to determine the four letters of your Myers-Briggs type.
For each pair of letters, choose the side that seems most natural to you, even if you don't agree with every description.

1. Are you outwardly or inwardly focused? If you:

• Could be described as talkative, outgoing • Like to be in a fast-paced environment • Tend to work out ideas with others, think out loud • Enjoy being the center of attention	• Could be described as reserved, private • Prefer a slower pace with time for contemplation • Tend to think things through inside your head • Would rather observe than be the center of attention
then you prefer **E** Extraversion	then you prefer **I** Introversion

3. How do you prefer to make decisions? If you:

• Make decisions in an impersonal way, using logical reasoning • Value justice, fairness • Enjoy finding the flaws in an argument • Could be described as reasonable, level-headed	• Base your decisions on personal values and how your actions affect others • Value harmony, forgiveness • Like to please others and point out the best in people • Could be described as warm, empathetic
then you prefer **T** Thinking	then you prefer **F** Feeling

2. How do you prefer to take in information? If you:

• Focus on the reality of how things are • Pay attention to concrete facts and details • Prefer ideas that have practical applications • Like to describe things in a specific, literal way	• Imagine the possibilities of how things could be • Notice the big picture, see how everything connects • Enjoy ideas and concepts for their own sake • Like to describe things in a figurative, poetic way
then you prefer **S** Sensing	then you prefer **N** Intuition

4. How do you prefer to live your outer life? If you:

• Prefer to have matters settled • Think rules and deadlines should be respected • Prefer to have detailed, step-by-step instructions • Make plans, want to know what you're getting into	• Prefer to leave your options open • See rules and deadlines as flexible • Like to improvise and make things up as you go • Are spontaneous, enjoy surprises and new situations
then you prefer **J** Judging	then you prefer **P** Perceiving

ISTJ Responsible, sincere, analytical, reserved, realistic, systematic. Hardworking and trustworthy with sound practical judgment.	**ISFJ** Warm, considerate, gentle, responsible, pragmatic, thorough. Devoted caretakers who enjoy being helpful to others.	**INFJ** Idealistic, organized, insightful, dependable, compassionate, gentle. Seek harmony and cooperation, enjoy intellectual stimulation.	**INTJ** Innovative, independent, strategic, logical, reserved, insightful. Driven by their own original ideas to achieve improvements.
ISTP Action-oriented, logical, analytical, spontaneous, reserved, independent. Enjoy adventure, skilled at understanding how mechanical things work.	**ISFP** Gentle, sensitive, nurturing, helpful, flexible, realistic. Seek to create a personal environment that is both beautiful and practical.	**INFP** Sensitive, creative, idealistic, perceptive, caring, loyal. Value inner harmony and personal growth, focus on dreams and possibilities.	**INTP** Intellectual, logical, precise, reserved, flexible, imaginative. Original thinkers who enjoy speculation and creative problem solving.
ESTP Outgoing, realistic, action-oriented, curious, versatile, spontaneous. Pragmatic problem solvers and skillful negotiators.	**ESFP** Playful, enthusiastic, friendly, spontaneous, tactful, flexible. Have strong common sense, enjoy helping people in tangible ways.	**ENFP** Enthusiastic, creative, spontaneous, optimistic, supportive, playful. Value inspiration, enjoy starting new projects, see potential in others.	**ENTP** Inventive, enthusiastic, strategic, enterprising, inquisitive, versatile. Enjoy new ideas and challenges, value inspiration.
ESTJ Efficient, outgoing, analytical, systematic, dependable, realistic. Like to run the show and get things done in an orderly fashion.	**ESFJ** Friendly, outgoing, reliable, conscientious, organized, practical. Seek to be helpful and please others, enjoy being active and productive.	**ENFJ** Caring, enthusiastic, idealistic, organized, diplomatic, responsible. Skilled communicators who value connection with people.	**ENTJ** Strategic, logical, efficient, outgoing, ambitious, independent. Effective organizers of people and long-range planners.

The Myers-Briggs Typology

twisty history. Its simplicity and popularity make it the ideal case study for how and why people classify themselves—a matter that becomes a great deal more important once those classifications get fed into computer algorithms. Tests like the MBTI, which offer strict division and regimentation of our self-identifications, are one way in which we make ourselves more comprehensible to computers.

In the 1940s, Katharine Cook Briggs and Isabel Briggs Myers started with four basic types, then leavened them with the personality theories of Carl Jung, as described in his 1921 book *Psychological Types*, in order to arrive at the four-axis classification. Jung's work is not as pseudoscientific as astrology, which is based on a geocentric model of the universe and requires Pluto to be a planet no matter what the scientists say. But there is little to suggest that Jung's work is rooted in anything beyond his casual observations, dubious intuitions, and free associations.* Briggs and Myers, who had little background in the social sciences, joined Jung's work with their own anecdotal experience to generate their schema. In the 1950s, psychologist David Keirsey combined the Myers-Briggs model with four "temperaments," to create the Keirsey Temperament Sorter. The two models differ in their details but share the same four axes and sixteen types. Most subsequent classifications have been derived from either one or both of them.

I first took an online Web 1.0 version of the test sometime in college and was not at all surprised to discover that I was a judgmental INTJ, also known as the Mastermind—or less charitably, the Crackpot. OkCupid's Brutally Honest Personality Test, a takeoff on the MBTI, has this to say about the INTJ:

> People hate you.
>
> I mean, you're pretty damn clever and you know it. You love to flaunt your potential. Heard the word "arrogant" lately? How about "jerk"? Or perhaps they only say that behind your back.

* George Makari's *Revolution in Mind: The Creation of Psychoanalysis* paints an unflattering portrait of the dawn of psychoanalysis, showing its pioneers to be ingenious but often undisciplined creative minds grasping in the dark toward the incomprehensibilities of the psyche, trading creation myths among one another in a farcical race to create a unified theory of the mind *ab ovo*. The result was a secular mythology of the mind whose greater contours remain with us today. Sigmund Freud, for all his intellectual caprice, still comes off as the sharpest mind of the lot, Alfred Adler as the most sensible. One psychoanalytic theorist was far more aggressively specious, racist, and arrogant than the others: Carl Jung.

> That's right. I know I can say this cause you're not going to cry. You're not exactly the most emotional person. You'd rather spend time with your theoretical questions and abstract theories than with other people.

This description makes for a needed complement to the ego-inflating verbiage usually employed. A sampling:

> INTJs are ambitious, self-confident, deliberate, long-range thinkers. INTJs are known as the "Systems Builders" of the types, perhaps in part because they possess the unusual trait combination of imagination and reliability. Whatever system an INTJ happens to be working on is for them the equivalent of a moral cause to an INFJ; both perfectionism and disregard for authority may come into play, as INTJs can be unsparing of both themselves and the others on the project.
>
> They generally withhold strong emotion and do not like to waste time with what they consider irrational social rituals. This may cause non-INTJs to perceive them as distant and reserved.

Read enough of this bombast and one will surely start displaying the arrogance of the Crackpot. The INTJ sounds like a pretty great thing to be, but all the types are described in similarly glowing rhetoric. As long as the rhetoric for my type is a little *more like me* than the descriptions of the other types, I have some reason to believe I'm an INTJ. Perhaps I do have more in common with Stephen Hawking and Isaac Newton (whom Keirsey.com has determined to be INTJs) than with Elvis Presley or Ronald Reagan (declared by Keirsey.com to be the exact opposite of my type, ESFPs), but I'm not quite ready to see a shared essence between Isaac, Stephen, and me.

Personality classifiers are not a purely descriptive process. We also prescribe our own type when embracing any such model, whether it's astrological signs, Myers-Briggs, or the Enneagram. If I lend credence to any of these models, then seeing myself as a Sagittarius, an INTJ, or a type 2 will grant me a point of reference and guide me in the future.*

* In the sidereal, astronomically correct zodiac, my birthdate falls under the short thirteenth sign of the Serpent Holder Ophiuchus, progenitor of the archetypal Greek healer Asclepius (linked to the Egyptian pharaoh and healer Imhotep), a sign that apparently classifies me as a

Cliff Johnson's 1987 puzzle game *The Fool's Errand*, based on the iconography of the tarot

Even telling people, "I think Myers-Briggs is nonsense, but in lieu of me describing my personality, I'll admit that the INTJ is vaguely in the ballpark" will condition them to see *me* in that way. If the categories are flexible (we can't change our birth month or astrological sign), I might doubt the classification, but simply *knowing* my classification will condition my experience. If the classification matches me reasonably well, I'll likely be biased to think my actions fall within that framework. My future actions might serve to reinforce the classification further. It may not be a bad thing to identify as an INTJ—it could even be healthy if it supports a positive self-image and encourages productive non-crackpot activity. But that's very different from believing it to be a scientific diagnosis, which it certainly is not.

I discovered the tarot deck through Cliff Johnson's wonderful and maddening 1987 computer puzzle game *The Fool's Errand*. Several years later

---

secretive mastermind with grand architectural plans and a distaste for authority and injustice, prone to being hypercritical, alienated, and impatient. I think astrology is nonsense, but as far as signs go, I could have done a lot worse. I also was born in the Chinese year of the Dragon—the arrogant yet humane Fire Dragon, specifically—and it's a shame these not-inappropriate signs should have been wasted on someone who doesn't believe in them.

The Hanged Man, the Hermit, and the Ace of Swords,
from a nineteenth-century Italian Minchiate deck

I read Italo Calvino's *The Castle of Crossed Destinies,* which uses a two-dimensional layout of cards to construct stories by reading sequences of cards across and down.* The mythology of the cards intrigued me—as myths, not as magic. So rather than having a reading done, I picked the cards that I felt were most appropriate for me. If I was going to let my fate be guided by a mythology, I wanted to guide how I applied the mythology to myself. So I decided on the enigmatic Hanged Man, the melancholic Hermit, and the truthseeking Ace of Swords.†

Is it cheating to pick tarot cards instead of drawing them at random? Only if you think the game itself is meaningful. I would argue that it's far healthier to choose one's cards instead of having them be dealt out from a shuffled deck. Classifications like the MBTI are closer to tarot than clinical diagnoses. They aren't as random as tarot readings nor as contrived as picking tarot cards yourself. They exist somewhere in between those two approaches, *not* as any sort of diagnostic test. MBTI tests usually offer a couple dozen questions that slot test-takers into one of the sixteen types, but the shorter ones just contain four questions,

* Calvino was a member of the Oulipo, and this was one of his experiments in writing stories based around formal constraints, much like Queneau's *Exercises in Style* and Perec's *Life: A User's Manual.*

† There is an old *Simpsons* episode in which a Tarot reader explains that Death signifies beneficial change, but gasps in horror when The Happy Squirrel appears.

one for each binary axis, allowing the test-taker to pick and choose. It's not a diagnosis, it's a selection.

In contrast, a clinical psychological test like the Minnesota Multiphasic Personality Inventory is constructed by normalizing around a sample population. In its classic incarnation, the MMPI-2, the MMPI offers over five hundred yes/no questions, including some duplicates, that assess the test-taker's personality on dozens of individual axes and around ten synthetic scales, including hypochondria, gender presentation, depression, and paranoia. It also includes "validity scales," which measure the honesty and consistency of the test-taker.

I took the MMPI when I was in my early teens and registered as normal on most scales save for one—depression. This came as a relief, as depression was one of the more innocuous scales next to mania, paranoia, schizophrenia, and the like. I was told I had been *very* honest on the test, which I suppose I had been. The test, I concluded, had been made for people exhibiting more severe indications of mental and emotional maladjustment than myself, and so I hadn't had much cause not to be honest in the first place. But the MMPI appeared empirical in a way that Myers-Briggs did not: it was assessing me relative to a chosen sample population that exhibited certain personality traits. That's not to say the MMPI is flawless: the normative component of such a test carries with it some serious complications, though far subtler than those of Myers-Briggs tests. Unlike the MMPI, Myers-Briggs never calibrated responses. It presents its dichotomies independent of any empirical basis and demands that you throw yourself into them.

The upshot is that Myers-Briggs and typologies like it are inherently self-classifications. To embrace a test like the Myers-Briggs is, in effect, to say, "Here is how I wish to see myself!" and "Here is how I want to be seen!" Anyone dissatisfied with their Myers-Briggs result should have little difficulty in skewing a subsequent test to obtain a more satisfying result. This is cheating only in the sense that it's cheating to look in the mirror to see if you missed a spot while shaving. As with astrology, the choice to use Myers-Briggs is more meaningful than any specific result. We care a lot about what classifications we put ourselves into.

Bitwise and Byte Foolish

> I will not be pushed, filed, stamped, indexed, briefed, debriefed, or numbered. My life is my own.
>
> —NO. 6, *The Prisoner*

I dwell on Myers-Briggs for three reasons. First, it's awfully popular. It or some variation of it is used in corporations and training across the country as a productivity tool and a means for workers to better understand themselves. Myers-Briggs shapes a substantial portion of social interactions. Second, Myers-Briggs serves as a simple and relatively nonloaded example of the sort of classifications that happen in far more controversial and high-stakes arenas: gender, race, class, sexuality, political affiliation, diseases and afflictions mental and physical, and criminality.

Finally, classifications like Myers-Briggs are extremely computer-friendly. If we accept the framework provided by the MBTI, I can describe my personality to you on the fingers of one hand, each finger representing one axis, extended or retracted. By seeing which fingers are extended, you will know who I am. In the Myers-Briggs schema, a person's personality type can be represented with a mere four bits of data; that is enviable compression for a description of a personality. (One's astrological sign requires four bits as well.) When computers are introduced, they work with these sorts of typologies to classify data. Here's how I could encode a person's Myers-Briggs type in C++:

```
struct {
  bool fExtrovert;
  bool fIntuitive;
  bool fThinking;
  bool fJudging;
}
```

Each Boolean variable above is set to either true or false. A true indicates that a person's type is that of the name of the variable. A false indicates that a person's type is the opposite of the name of the variable. It's only four bits of information, but each variable occupies at least one entire byte (a byte is almost always eight bits long, and the smallest, atomic unit of computer storage), so this is at least thirty-two bits of storage here. I can improve on this by using a single byte-length variable (the one-byte `char`, the size necessary to store a single alphanumeric character), with each particular bit designating one of the Boolean flags. It would look like this:

```
#define MBTI_EXTROVERT      0x01 // (or 1)
#define MBTI_SENSING        0x02 // (or 1 << 1)
#define MBTI_THINKING       0x04 // (or 1 << 2)
#define MBTI_JUDGING        0x08 // (or 1 << 3)

unsigned char MBTIType;
```

So for an ESTJ, all the bits are on, yielding the binary number 1111. For an INFP, all the bits are off, yielding binary 0000 (or just plain 0). For my type INTJ, the last two bits are on and the first two are off: 0011. The variable **`MBTIType`** still has bits left unused after the first four, but we'll leave that be.

If I want to assign and read the type in one go, I could enumerate all sixteen types and assign them consecutive numbers beginning with 0, all the way up to 15. This is aptly called an enumerator:

```
enum MBTIType {
  INFP = 0,
  INFJ = 1,
  INTP = 2,
  INTJ = 3,
  ISFP = 4,
  ISFJ = 5,
  ISTP = 6,
  ISTJ = 7,
  ENFP = 8,
  ENFJ = 9,
  ENTP = 10,
  ENTJ = 11,
  ESFP = 12,
  ESFJ = 13,
  ESTP = 14,
  ESTJ = 15
};
```

What you may not have noticed—and what computers certainly *would* notice—is that these last two representations, the enumerator and the four-bit variable, are identical. I have ordered the enumerated type such that the value of any given MBTI type is the same as that which you would get by putting the respective bits together in the four-bit representation. The enumerator contains four bits of information. INFP is 0, just as it was when all four bits were off in the preceding example and we obtained binary 0000. ESTJ is 15, which translates to binary 1111. (Instead of powers of 10 for places, as in decimal, binary uses powers of two, the first four being 1, 2, 4, and 8. 1 + 2 + 4 + 8 = 15.) INTJ is 3, which translates to binary 0011 (1 + 2).

For the computer, these representations are identical, because in machine language, the program works exclusively with numbers. The code that I write to compare the value of an MBTI variable against the code for INTJ is comparing that variable against a value of 3 (or 0011, more accurately, since it's all just 1s and 0s at the bottom). Whatever meaning I gave it, *even in the code,* has disappeared by the time the CPU is executing the code. Numbers are truly meaningless to computers.

The reduction of all data to numbers (bits, specifically) makes the computer a truly neutral arbiter of data. While we may enjoy playing around with personality types and going to corporate retreats, there's something far more unsettling about seeing a personality type coded into a fixed computational representation and having that representation stored indefinitely on a centralized server. "Hello, I'm an INTJ!" doesn't carry with it the sense of being indelibly marked, but having a centralized database (whether corporate or governmental) store a 3 designating me as an INTJ certainly does. The computer does not understand what it means to be an INTJ, but we do. And once that 3 is in there, the issues of who controls the value, who sees it, who gives it meaning, and how it is used become significant far beyond the limits of the machine's memory. Switching up my Myers-Briggs type should not be as difficult a process as changing my legal name, but as we lose personal control of our data, we also lose part of our ability to define and redefine ourselves.

I went along with the color-coding at the corporate retreat. I put on a green shirt and accepted my lot, but when it came time to form teams of five or six people for competitive activities (build the tallest tower out of office supplies, give accurate navigation instructions to a

5

SELF-APPROXIMATIONS

The Big Five (or Six)

> Putting facts into nice cleanly demarcated buckets of explanation has its advantages—for example, it can help you remember facts better. But it can wreak havoc on your ability to think about those facts.
>
> —ROBERT M. SAPOLSKY

MYERS-BRIGGS IS RECREATIONAL. The MMPI is clinical. What if there was a model that claimed to capture the entirety of a person's personality, not just a few aspects of it? There already are several. Somewhere in between the MMPI and Myers-Briggs are popular yet scholarship-backed taxonomies like the "Big Five" personality traits. The "Big Five" has achieved prominence over the past twenty-five years, after a number of psychology researchers reached consensus on the general model in the 1980s. The model clusters personality traits under five umbrella terms: Neuroticism, Agreeableness, Conscientiousness, Extroversion, and Openness (which produce the acronyms OCEAN or CANOE, depending on one's preference for seeing the psyche as a murky depth or a leaky boat). Each is a linear spectrum.

The best-known variant of the family of Big Five models is the NEO Five-Factor Inventory, created by psychologists Paul Costa and Rob-

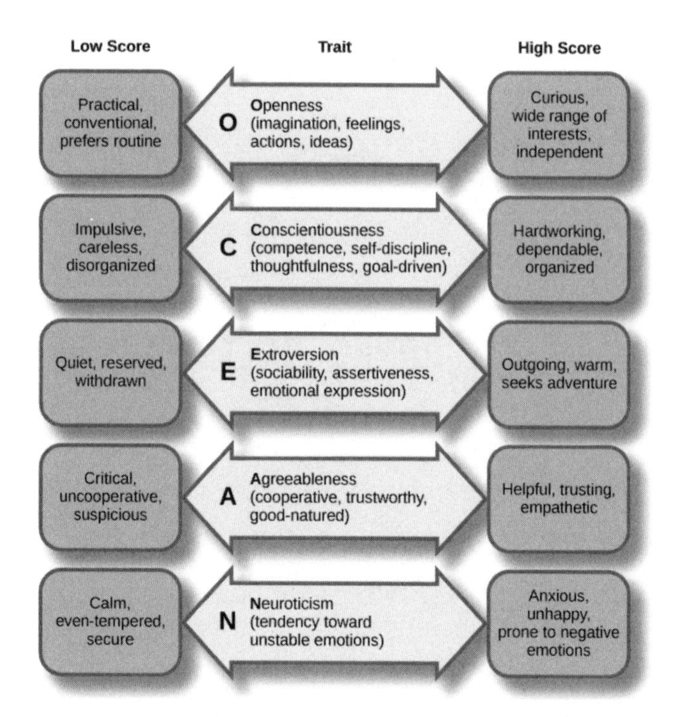

The Big Five personality traits

ert McCrae.* Costa and McCrae claim the Big Five model is cross-culturally universal *and* reflects basic dispositional tendencies that are predominantly fixed by genetics at birth, though both claims have been disputed. In his popular introduction *Personality,* Daniel Nettle deems the Big Five to be the master model of personality: "Most constructs that had previously been measured can actually be subsumed under the five-factor framework. . . . It is very likely [the Big Five] will turn out to be shorthand for suites of differences in neural structure and function across multiple brain regions." This audacious claim—that we have discovered *the* neurobiological structure of human personality—has also been disputed.

Two other psychologists, Kibeom Lee and Michael C. Ashton, have created a six-factor model, HEXACO, that takes OCEAN's Big Five and adds a sixth, purportedly independent axis: Honesty vs. Humil-

* Other strong proponents include psychologists John Digman and Lewis Goldberg.

FACTOR	ADJECTIVES
Honesty–Humility	Sincere, honest, faithful/loyal, modest, unassuming, fair-minded versus sly, deceitful, greedy, pretentious, hypocritical, boastful, pompous
Emotionality	Emotional, oversensitive, sentimental, fearful, anxious, vulnerable, versus brave, tough, independent, self-assured, stable
Extroversion	Outgoing, lively, extroverted, sociable, talkative, cheerful, active versus shy, passive, withdrawn, introverted, quiet, reserved
Agreeableness	Patient, tolerant, peaceful, mild, agreeable, lenient, gentle versus ill-tempered, quarrelsome, stubborn, choleric
Conscientiousness	Organized, disciplined, diligent, careful, thorough, precise versus sloppy, negligent, reckless, lazy, irresponsible, absent-minded
Intellect/Imagination/ Unconventionality	Intellectual, creative, unconventional, innovative, ironic versus shallow, unimaginative, conventional
* Intellect/Imagination/Unconventionality is sometimes known as Openness to Experience.	

The six-factor HEXACO personality model

ity. In this scheme, the Big Five's Neuroticism has been renamed Emotionality, while Openness is also labeled Intellect/Imagination/ Unconventionality. (Extroversion becomes the X in HEXACO.) The authors of Big Five systems are, unsurprisingly, dismissive of HEXACO. McCrae downplays Honesty-Humility as a superfluous correlate of Agreeableness.*

Personality models like the Big Five and HEXACO promise neat, authoritative classifications of our personalities, more scientific than Myers-Briggs and broader than the MMPI. Both are well suited to computational classification. The problem is which model to pick. McCrae and Costa boldly say, "The existence of these five factors is simply an empirical fact, like the fact that there are seven continents on the earth," but scientific rigor does not permit that conclusion, any more than we

* "Ashton's model basically divides FFM A[greeableness] into two factors, the second called Honesty-Humility."

can say that the four humors are an empirical fact. Such things are heuristic constructs. With the Big Five and HEXACO, we've been given two yardsticks, each with different sets of markings on it, and asked to decide which yardstick has the correct markings. *There is no correct answer*. We are not measuring an objective phenomenon, but rather trying to measure the subjective measures *themselves*. Each model draws attention to certain personality distinctions while overshadowing distinctions that the model cannot accommodate. The personality models, while not arbitrary, are *selective* and *inadequate*. OCEAN and HEXACO proponents each point to behavioral studies supporting their respective views, but until we are able to determine personality from a person's genes and neurobiology, science cannot choose one over the other. Yet the dispute will be "resolved" when one model proves more popular than the other—or, more likely, when some third model supplants both.

What if our OCEAN (or HEXACO) types were associated with our Facebook profiles, our corporate HR records, our government files, and so on? In fact, OCEAN and HEXACO are used by corporate training organizations, just as Myers-Briggs typologies are. Would we want to be placed into a HR database indicating we've scored low on Agreeableness? Once we start using one system, like OCEAN, there are high costs to transitioning to another like HEXACO. Faced with the burden of having to *reclassify* people, there will be a strong incentive to believe that the existing classification is indeed the correct one.

In practice, we need taxonomies to provide *some* coarse classification of types. Humans speak constantly in reductive generalizations, because we lack the physical and mental capacities to process the world in all its raw detail. The danger lies in adopting and privileging a *single* taxonomy. When we do that, we run the risk of stomping on the complexities that slip through the sieve of whichever conceptual system we use. And the more we hew to one system, the less likely we are to revise that system or transition to a new and possibly better one. No four-, five-, or six-factor system will ever be adequate for the complexities of human psychology, but a new system is often preferable to the existing one, if only because it preserves variety and reminds us that our current system is hardly the be-all and end-all.

Diagnostics and Statistics

> It would be indeed unusual if it turned out that the set of orders that our mind is able to construct and accept, having as it does a deep sense of "understanding the essence of things," matches precisely the set of all possible orders to be detected in the Universe as a whole. We should admit that this is not impossible, yet it does seem highly improbable.
>
> —STANISLAW LEM

MY PARENTS had three huge bookshelves along the wall of our living room, and as I grew up I made my way through their collection. There was a good dose of golden-age science fiction, Carl Barks's Donald Duck comics, the *World Book Encyclopedia,* Max Shulman's Dobie Gillis stories, and Martin Gardner's annotated Lewis Carroll. There were also the mid-century books for adults I didn't read until much later: *Serpico, Catch-22,* and Moss Hart's *Act One,* alongside psychiatric texts like Freud's *Totem and Taboo,* a title I found creepy. But the most perplexing book they had was one I also spent a great deal of time with: the *Diagnostic and Statistical Manual of Mental Disorders, 3rd Edition, Revised* (the *DSM-III-R* for short), the psychiatrist's bible, published by the American Psychiatric Association. This 1987 reference book, which was over five hundred pages of dense type, contained an elaborate numerical taxonomy of *every* mental disorder under the sun, from alcoholism to schizophrenia to anorexia, along with detailed criteria for diagnosing whether a patient possessed each disorder. I say *every* mental disorder because, as I learned, psychiatric practice often required that all patients be labeled with one or more of the labels in the *DSM.*

Both of my parents are psychiatrists, and the 1980s were an exciting and transformative time for psychiatry for two main reasons: the introduction of the new wave of SSRI antidepressants such as Prozac and Zoloft, and the diagnostic revolution of the *DSM-III* (1980) and its successor, the *DSM-III-R* (1987). As a child, I witnessed two manifestations of these revolutions in our house: the preponderance of Prozac- and Zoloft-branded paraphernalia like pens, mousepads, Post-it note cubes, and penlights, and the constant presence of the intense blue cover of the *DSM-III-R*.

By the late eighties, my family owned an IBM-compatible computer (a Compaq, specifically) as well as an Epson dot-matrix printer.* My parents were not good with computers, so they enlisted me to help format, spell-check, and print their psychiatric reports for them. We used WordPerfect 5.1, a solid if austere word processing program that presented little more than a blank blue screen on which to type, along with a set of function key shortcuts that led to menus and other various options.† The blue screen matched the blue of the cover of the *DSM-III-R*.

My father worked for the county health system but occasionally did consultation work for the state in assessing disability candidates. The state provided no information on the candidates, other than to let him interview them a single time to provide input into the state's determination of whether or not the candidate was entitled to disability benefits. Given this suboptimal clinical situation, my father tried to be charitable, if cautiously so. The state was not inclined to be charitable. The reports couldn't be conclusive in any way under the circumstances, but the state required them before they would pay disability. And a big part of the report was the *DSM* diagnosis.

My father typed his reports, his typewriter-trained fingers loudly hammering the keys, then, after blanking all identifying information, he asked me to work my computer magic. I would then underline and bold the appropriate parts, align everything properly, leave room for his signature, and print out the final product. These consultation reports contained summaries of the patients' backgrounds, complaints, and cognitive assessments, such as whether they could recite the alphabet forward and backward, spell simple words correctly, name the first and last five presidents, etc. At the end was the terse *Prognosis,* which

* Anyone who ever owned one of the classic Epson printers will remember how loud they were, the banshee screech of each printed line coming out like laser fire. A line of underscores (_______) had a thinner, more monotone sound than the high-pitched chatter of letters and numbers. The English composer, improviser, and music maker Hugh Davies took advantage of the sonic variability to construct dot-matrix printer compositions in the 1980s.

† Some remain with me via muscle memory: F5 brought up the file system interface. F10 saved the current document. F7 (followed by an "N" and a "Y") exited the program. F2 was the spell-checker. There was a thin, long strip of function key shortcuts that came with WordPerfect and sat above the function key row of our keyboard for years.

was frequently "good" or "fair" but sometimes "guarded" (which, as my father explained to me, was the opposite of "good"). In between the summary and the prognosis was the all-important diagnosis, consisting of five "axes," as defined by the *DSM*. Here they are, roughly, before they were eliminated in *DSM-5:*

Axis I: The diagnostic code for the principal disorder, as defined by the *DSM*.
Axis II: Personality disorders and intellectual disabilities.
Axis III: Relevant non-psychiatric medical conditions.
Axis IV: Relevant psychosocial, environmental, and/or situational factors.
Axis V: Global Assessment of Functioning (GAF) score.

The first and last were by far the most important, to the extent that my father often left the middle axes blank. This was the first time I'd witnessed human behavior classified in such a concrete and discrete fashion. As a budding quant, I was fascinated by the strangeness of it. People I did not know were being described in a clinical language that didn't match anything of my understanding of people. This clinical language also seemed to describe a private world to which I didn't have access. I learned words like "anxiety" and "borderline" and "obsessive-compulsive,"* but none of them appeared to apply to human behavior as I witnessed it. It was a foreign dialect.

Axis V, the Global Assessment of Functioning, seemed simplistic to me even then. It was a scale from 1 to 100, where 100 was "superior functioning," 50 was "serious impairment in functioning," and anything under 20 signified danger to one's self or others. The GAF struck me as a bit . . . general. But then, that appeared to be its role: a rough measure of how badly off patients were and thus how in need of treatment they were. I asked my father how he could possibly differentiate between a 55 and a 56, and he told me that he couldn't, and he didn't. No one could.

Axis I was much more complicated and much more interesting. It

* In fact, I'd learned "anxiety" from *Peanuts,* specifically from Charlie Brown's line "My anxieties have anxiety." Lucy came to symbolize many Americans' ambivalent image of psychiatry for many years. My father, however, bemoaned how poorly psychiatrists were portrayed in popular culture, from Dr. Mabuse to Hannibal Lecter.

was an elaborate list of coded diagnoses, usually identified by three digits and two decimals, which elaborated every possible psychiatric diagnosis that could go into a patient's file. Post-traumatic stress disorder, bipolar, paranoid schizophrenia, borderline personality disorder; just as these terms had been brought into common parlance in the 1980s through usage of the *DSM-III,* they became known to me through my parents' use of them as codes. It was their definitions that were odd.

Every disorder had a number, but the numbers were misleading. I had read Martin Gardner's polemics against pseudosciences like Scientology and Velikovskyism, and I was on the lookout for more. I liked math, but the *DSM* wasn't math. It was just *numbers.* Catatonic schizophrenia did not become a more objective label simply by assigning it the number 295.2. The *DSM* notoriously uses a criterial system to determine diagnoses: one need not possess *all* the markers of a disorder to be diagnosed with it, only *enough* of them. Many disorders entail having "5 of the following 7" or "at least 4 of the following 10" behaviors or tendencies listed.

This did not fit any conception I had of what a disease was, nor did it fit psychiatry as my parents had described it to me. More importantly, it did not even fit any conception I had of what a *taxonomy* was.* My father was quite a taxonomizer himself—of television, coins, comic books, and science fiction—and *those* taxonomies made far more intuitive sense to me than the *DSM*. These checklists didn't function well as overall rubrics; some of them, at least to my preadolescent self, seemed random. I was hardly a good judge of personality at that point, and the way symptoms and behaviors were arranged and listed seemed puzzling, if not entirely arbitrary. I presumed that there was some expert logic that explained why it was *these* diseases that people had or did not have.

When I started hearing people employing terms like "OCD" and "manic depressive" in daily life, they were invoked with little regard to the bloodless lists of criteria given in the *DSM*. The *DSM* categories became prescriptive: they *generated* matching experiences and

* The term of art for classification of diseases is "nosology," which I see no need to use any more than one *needs* to use the correct collective noun for every sort of bird. The difference between "a murder of crows" and "a group of crows" is the difference between an in-group and an out-group's vocabulary. So too with the difference between "nosology" and "taxonomy" or "classification."

symptoms as often as they diagnosed them. Freudian psychoanalysis had yielded symbolically sexualized dream imagery, a psychodynamic model of the unconscious at war with itself, and a powerful account of neurosis based in childhood, adolescent, and adult sexuality, usually in some deformed or defective state. We carry so many of these early psychoanalytic concepts with us unwittingly today—the superego, repression, Freudian slips, the unconscious itself—that they have become inextricable from our culture, even if Freud's all-consuming vision of sexual neurosis has faded. The legacy of the *DSM* is harder to gauge. Where psychoanalysis *aestheticized* and *dramatized* mental illness for the general public, the *DSM classified* and *quantified* it. This is, loosely, the shift from qualification to quantification. We moved from *describing* one's *mishigas* to *labeling* it.*

Sometimes I would look up a disorder and wonder how close I was to it, since many disorders possessed at least one sufficiently innocuous criterion that I matched. One of schizophrenia's criteria was "disorganized speech." Did that mean I might be schizophrenic because I rambled on with incoherent sentences sometimes? "Lack of enjoyment of experiences" was listed on *many* disorders! How close was I to madness? Not very, according to my parents. They explained that clinical assessments were not made by such criteria, but by the reasonable judgment of adults like themselves who wouldn't diagnose people as crazy just because they ticked off the right number of boxes under a *DSM* category. So what, I asked, was the purpose of these categories? Getting paid, they told me.

Years later I had my own perverse clinical encounter with the *DSM*. When I was twenty-six years old and on vacation by myself in New York City, I had a panic attack—maybe. I had spent the evening first at an Evan Parker concert uptown, then at a Keith Rowe and Toshimaru Nakamura concert downtown. In between was a breakneck car ride down the streets of Manhattan by a very high-octane friend, Pei-Yi, while I and my other friend Chris held on to the armrests. After the second show, Chris and I had some very bad chicken—undercooked

* A label is not a fixed abbreviated description. A label has many conflicting definitions, each of which is no more than a grouping of overlapping labels itself, and there is no adjudicating process for determining which one applies in a particular circumstance. Labels are extremely fuzzy and circularly interrelated entities. There is little that cannot be thrown into confusion by taking a particular term, whether it's "borderline" or "mania" or even "sick," and asking, "What do you mean by that?"

and sickly pink—at a Chinese restaurant near Tonic, the much-missed club where Rowe and Nakamura had played. Chris headed uptown to his apartment, and I went back to my hotel. Happy and exhausted, I lay down on the bed to unwind, and then my body went mad.

Time slows down in unfamiliar and frightening situations: just before a car accident, the couple seconds of seeing a car heading far too close to me in the rearview mirror stretched out to feel like minutes. What happened on that night unfolded over what seemed like hours. An unfamiliar wave of discomfort and numbness passed through my body. In my arms and legs and particularly in my head, I felt a terrifying and *physical* sense of dislocation, as though I were losing all sensation, except that instead of numbness, it was a new kind of pain that I had never experienced before. It was not a sharp pain or a dull pain, not a throbbing pain or a stabbing pain, not tension or heat or cold. I was still very much in my body, but my body had become a far more hostile place. It was pain without being pain, a sensation that didn't feel comparable to anything I had known but that was incredibly unpleasant and persistent. My head swirled. I could not understand what was going on, and I was all alone in New York City. I called 911.

I remember being in an ambulance, telling the attendant that I felt like I was going to pass out. He told me that people who say they're going to pass out never pass out, which was somewhat reassuring but didn't stop me from thinking that I was dying. I called Chris on the phone and told him that there might have been something wrong with the chicken and that I was going to the hospital. I left messages with my wife and my mother. The feeling continued to roil me without subsiding or worsening. I stayed alert, wondering what was coming next. I was panicked, but not irrational. I was rigidly focused on the situation at hand, trying to figure out whom to contact and what to say. I had made sure to tell my wife and mother that I loved them. I reflected that it would be a shame to die at this moment, because things had been going well recently and I was making progress in my life.

Sometime later, I was lying on a stretcher in an emergency room. The feeling had slightly abated. I still felt terrible, but it was harder to differentiate between the initial, awful feeling and the subsequent terror of dying that had joined it shortly after. I began to think that I was going to live. The nurse did an EEG on me and took other vital signs—all were normal. Chris showed up and told me that on hearing my mes-

The Temptation of St. Anthony, by Pieter van der Heyden, after Bruegel

sage, he'd panicked and stuck a toothbrush down his throat to induce vomiting.

"What did I sound like?" I said.

"You sounded like you were dying."

I said that I now didn't think that I was. When the on-call doctor arrived, he told me that it was a panic attack, which under the circumstances was about the best-case scenario. I spoke to my wife, who was relieved to hear that I was alive, and my mother, who told me that it did indeed sound like a panic attack. There was nothing I had been panicked *about,* but she replied that that was not how these things worked. Panic was a state of being, not a state of mind. Chris went with me back to the hotel and stayed with me overnight. I felt light-headed and strange in the morning, as though my head had been depressurized, but the terrible sensation had otherwise mostly passed.

What seized me that night still defies easy description. The best visual portrayal of the sensation I know is in Pieter van der Heyden's print after Bruegel, *The Temptation of Saint Anthony*. An oversize head lies in a river, dragged down by chaos and drowning in surreal torment. My vision was unaltered, but the print captures the sensory corruption and loss of bodily autonomy that I suffered. The world remains unchanged; it is just one's apprehension of it that has gone horribly wrong. The ordinary, passive act of *living* and *experiencing* becomes painful.

What had happened to me? The *DSM-IV-TR,* which was the latest edition at the time, gives the criteria for a panic attack as consisting of four or more of the following symptoms.* I have bolded the symptoms I experienced on that night.

1. **palpitations, pounding heart, or accelerated heart rate**
2. sweating
3. trembling or shaking
4. sensations of shortness of breath or smothering
5. feeling of choking
6. chest pain or discomfort
7. nausea or abdominal distress
8. **feeling dizzy, unsteady, lightheaded, or faint**
9. feelings of unreality (derealization) or being detached from oneself (depersonalization)
10. fear of losing control or going crazy
11. **fear of dying**
12. **numbness or tingling sensations (paresthesias)**
13. **chills or hot flushes**

At five symptoms, my experience qualified as a panic attack—barely. I was dissatisfied with the diagnosis, however, because the symptoms befell me in two stages. There was first the initial terrible feeling, which I'll call *nerve corruption,* as my body was reporting mysterious and awful sensations to my mind that were foreign to me—and apparently inaccurate. The feeling of nerve corruption contained, at best, the last two symptoms: numbness and chills. Everything else followed sometime later and was pretty obviously the product of my worrying about nerve corruption, not something that was intrinsically part of the nerve corruption. So was the nerve corruption itself a panic attack, or had it caused a panic attack?

A few weeks later, the nerve corruption descended upon me again.

* The *DSM-5* reorders the symptoms to put the two fears last, but otherwise leaves the list from the *DSM-IV-TR* as is. It also puts "going crazy" into quotes. It adds, "Culture-specific symptoms (e.g., tinnitus, neck soreness, headache, uncontrollable screaming or crying) may be seen. Such symptoms should not count as one of the four required symptoms." I don't see how those symptoms are culture-specific, but I am not a psychiatrist.

A mytho-symbolic portrayal of the epileptic's relation to the world, from David B.'s *Epileptic*

This time, however, I was certain that as awful as it felt, I was not dying and was probably not in any real danger either. I sat down in a chair, closed my eyes, and breathed slowly, trying to track this strange *sensation**—as it oscillated and twisted inside my head of its own accord. After twenty terrible minutes, the nerve corruption slowly abated. That time there had not even been a chill, possibly owing to my being in a warmer room than I had been in in New York. There had been a tingling half-numbness alongside the indescribable nerve corruption. Chest pain? No. Hyperventilation? No. Racing thoughts? No. I had, in fact, made a point to stay as calm as possible during the nerve corruption so as to observe its essential elements.† I continued to experience

* What is the difference between a sensation and an emotion? French conflates the two in the word *sensibilité,* which is sensation rather than sensibility per se, and since everything is neurochemistry anyway, the distinction appears to be more by convention than anything else. Is pain an emotion? I can distinguish the heavy, pillowy weight of sadness and despair from the spiky intrusion of guilt, but both are a great distance from the malignant, irresistible, and nonnegotiable steamroller assault of biochemical depression.

† I had spent some time reading about the Buddhist philosopher Nagarjuna and his philosophy of emptiness, and the Buddhists' radically contingent view of phenomenal experience came as a great aid to me by helping me stand at a remove from the bizarre things my brain and body were doing to me. It also had the less salutary effect of teaching me how tenuously each of us is connected to an ordinary, manageable experience of the world, such that we stand, by the grace of our bodies, only a few steps away from unbearable agony or mad-

episodes like this infrequently over the next few months, until they gradually dissipated. They usually came in the evenings or at night, sometimes when I slept, very rarely when I was active or at work. If they had any correlation to stressful events in my life, I could not identify the pattern.

Here is what a psychologist told me: the New York episode was a panic attack, and the *only* panic attack I had suffered. I protested: Why should the fact that I thought I was dying the first time make it somehow different? I fully understand that working myself into a frenzy over a weird and unpleasant feeling will cause me to panic, I told him, but I'm not doing that anymore. I do not recall getting a satisfying answer. So instead of panic attacks, I was experiencing something that fell between the cracks of the *DSM*'s categories. Short of having an experience of nerve corruption while hooked up to assorted monitoring equipment, I expect it will remain a mystery to me and to science. My mother once proposed it might have been an unusual form of migraine, which gained some support when I found out that the frequent minor headaches I'd suffered my whole life were in fact migraines, despite lacking aura or the more severe features frequently associated with migraines. But the nerve corruption didn't share any apparent link to anything I'd thought of as a headache either, which was a far more familiar form of pain. I found later that I could diminish the length of a nerve corruption episode by taking a small dose of a tranquilizer, which indicated panic may have been the cause again—except that, according to the *DSM*, I was not actually having panic attacks.

The inability to show people my inner experience has never been more frustrating than in this case. I thought that the fundamental criteria for the diagnosis were ridiculous. I thought the doctor's adherence to them was ridiculous. Yet the ostensible clarity granted by the numbers and the criteria undeniably give *DSM*-based diagnoses a greater veneer of scientific objectivity. And indeed, this was the problem that the *DSM-III* was created to solve. The *DSM* hadn't interfered with the treatment of my nerve corruption; it just failed to taxonomize

---

ness. David B.'s graphic novel *Epileptic* is a harrowing and hermetic portrayal of that thread of coherence being severed. The psychologist Louis Sass has written perceptively on these issues, and the distortions through which language puts them, in *The Paradoxes of Delusion: Wittgenstein, Schreber, and the Schizophrenic Mind* and *Madness and Modernism: Insanity in the Light of Modern Art, Literature, and Thought.*

it.* Psychiatry's greatest contribution had been to *humanize* the mentally ill and secure for them somewhat better treatment than they had been allotted in the past. This was a slow process, with many missteps and horrors, but the degree of sympathy we extend to the mentally ill is far greater now than it was in the pre-Freudian era. Nonetheless, there is little of the theoretical work in Freud and his followers—what Joyce called "Jungfraud's Messongebook" in *Finnegans Wake*—that stands up to scientific scrutiny.

As psychotherapy became mainstream and institutionalized in the United States, medical authorities criticized it for lacking *diagnoses*. Certainly there were extreme forms of delusional schizophrenia that fit a reasonably consistent pattern, but for more minor ailments, the psychotherapeutic approach offered few ways to summarize what precisely was wrong with a patient and couldn't guarantee that different psychiatrists would agree on a diagnosis. The *DSM-II* notoriously classified homosexuality as a mental disorder, listing it alongside other "Sexual Deviations" such as sadism, masochism, and transvestitism. Other artifacts included "neurasthenia," "psychotic depressive reaction," and "involutional melancholia." The homosexuality classification was rectified in 1973, thanks to an effort led by the quantitatively oriented psychiatrist Robert Spitzer, but it was increasingly clear that many of the other categories, while less controversial, were terminally vague, overlapping, and sometimes incoherent.

Spitzer went on to be the central figure behind the full-bore rewrite of the *DSM-II* into the *DSM-III*, which was published in 1980. The *DSM-III* cleaned up the terminology of the *DSM-II* and separated

* The brief history of the *DSM* I give here draws heavily on the writings of Donald Goodwin, Samuel Guze, Allen Frances, Gary Greenberg, Michael Alan Taylor, Edward Shorter, and David Healy. With the exceptions of Goodwin and Guze, who anticipated and influenced the *DSM-III*, and Frances, who was the architect of the *DSM-IV* before becoming one of the preeminent critics of the *DSM-5* (alongside Robert Spitzer himself), they all think poorly of the *DSM*. In pursuit of balance, I read the defenses of the *DSM* methodology offered by people such as *DSM-5* architect Darrel Regier, William Carpenter, and Helena Kraemer. To my inexpert eye, they mostly fail to rebut the most serious charges of their critics, namely with regard to the fundamental lack of validity of the existing *DSM* taxonomy. I do not mean to litigate that dispute. Disputes on whether medications are being overprescribed will not be solved by appeals to unsound taxonomies—*or by attacks on them*. If a patient feels better after taking Prozac, the question is not whether he or she is mentally ill, but whether the prescription is responsible and safe. Since historical accounts of the *DSM* have been authored overwhelmingly by its critics, I have attempted to correct for that by avoiding the more contentious and subjective aspects of their criticisms where possible.

mental illnesses out into more logical and hierarchical categories, yet the *DSM-III* still was based on the taxonomical foundation of the *DSM-II*. Indeed, the *DSM-II*'s overall division of psychiatric illnesses into affective syndromes like depression, psychoses like schizophrenia, and personality disorders like borderline remained in the *DSM-III*. The revolution of the *DSM-III* lay instead in the introduction of a *criterial approach to diagnosis*. The *DSM-II* provided only a general description of each syndrome, so that the individual physician judged whether a patient fit it or not. The *DSM-III* did not require psychiatrists to match those generalized descriptions to patients. Instead, diagnosis was to be obtained through behavioral markers, the "5 out of 7" approach I mentioned above. Whether or not this produced a better taxonomy of actual diseases, it *did* accomplish something significantly more important, which was to allow for replicability and to explain a lack of replicability. If two psychiatrists disagreed on a patient's diagnosis, they could simply go down the checklist for each diagnosis and see which markers were met. Just as significantly, the *DSM-III* allowed for quantitative research to be performed on groups of patients. If a psychiatrist wanted to see the effect of a treatment on a particular disease, he or she could study patients diagnosed with *that* particular *DSM-III* classification. This proved crucial for the pharmacological boom ushered in by Prozac in 1986. The FDA required drugs to be prescribed for a specified set of diagnoses, so the increased standardization of the *DSM-III* enabled psychiatric drugs to enter the mainstream. Diagnoses were now much easier to fit not just onto severely mentally ill patients, but also higher-functioning people.

This transition bolstered the shift from *psychoanalysis* to *psychopharmacology*. The individual, unpredictable, and inexorable course of analysis, with its personalized, irreproducible hours of dialogue, was ill suited for any sort of standardized, quantifiable treatment regimen, and therefore a very bad match for the increasingly actuarial health insurance industry. For those wealthy enough to pay out of pocket, psychoanalysis remained relevant, particularly in places like New York where it was nearly a cultural signifier.* For everyone else, though, the *DSM-III* enabled a new sort of psychiatric treatment that did not equate

* New York is the only city I have lived in where people talk openly about their therapists and their relationships with them.

mental illness with severe pathology or Freudian analyses. Everyday neurosis could now be quantified and treated as one would treat any other physiological symptom.

For all the alarm bells that have rung about overdiagnosis and the inflation of mental illness, the problem is not with increasing the number of diagnoses per se, but with how society then deals with that greater pool of "mentally ill" people. And here both doctors and drug companies have a lot to answer for, as *DSM-IV* architect Allen Frances says:

> The business model of the pharmaceutical industry depends on extending the realm of illness. . . . The real problems begin at the intersection between *DSM* diagnoses and FDA drug indications. Once a drug is approved for a more popular *DSM* diagnosis, it sells much more. This has caused a vicious feedback loop, as newfangled drugs are pushed toward treating the most popular diagnoses. As the most popular diagnoses tend to be by nature the more mild diagnoses, this results in a trend toward overtreatment.

The slippery nature of the *DSM* categories exacerbated this feedback loop. Because they were not based in anatomical or neurophysiological reality, but were rather folk categories inherited from years of vague consensus among psychiatrists, these diagnoses were not set in stone in the way that, say, heart ailments are. Peer pressure from colleagues and financial pressure from drug companies could shunt these categories in directions that were not in patients' best interests. This played out disastrously in the case of childhood bipolar disorder, a new diagnosis meant to supplant some existing diagnoses of ADHD. Harvard professor Marcia Angell tells the story:

> Take the case of Dr. Joseph L. Biederman, professor of psychiatry at Harvard Medical School and chief of pediatric psychopharmacology at Harvard's Massachusetts General Hospital. Thanks largely to him, children as young as two years old are now being diagnosed with bipolar disorder and treated with a cocktail of powerful drugs, many of which were not approved by the Food and Drug Administration (FDA) for that purpose and none of which were approved for children below ten years of age.

> Biederman's own studies of the drugs he advocates to treat childhood bipolar disorder were, as *The New York Times* summarized the opinions of its expert sources, "so small and loosely designed that they were largely inconclusive."

The scandal broke out, as Angell tells it, when it became known that Biederman had received $1.6 million from drug companies, including those that made the very same drugs prescribed to those diagnosed with childhood bipolar. Instead of diagnoses driving the prescriptions, the prescriptions were driving the diagnoses. Psychiatric labels had been prodded in one particular direction—a more profitable one.

Machine Psychiatry

> [Freud] has not given a scientific explanation of the ancient myth. What he has done is to propound a new myth.
>
> —LUDWIG WITTGENSTEIN

The discrete taxonomies of the *DSM-III* were quite amenable to computers, and the desire to increase quantification of mental illness guided (and misguided) the latest revision, the *DSM-5*. Frances, in his book *Saving Normal,* describes trying to refine the existing categories in the *DSM* without letting special interests corrupt them.

> We saw *DSM-IV* as a guidebook, not a bible—a collection of temporarily useful diagnostic constructs, not a catalog of "real" diseases. We tried to make this abundantly clear in the introduction to *DSM-IV* and at greater length in the *DSM-IV* Guidebook. Unfortunately, I am not sure anyone ever reads the introduction, and I know that few people have read the Guidebook.

If the *DSM* is not a catalogue of real diseases, you wouldn't know that from the extent to which categories like bipolar and borderline are thrown around clinically and the extent to which they are researched *as though* they are real. The computers processing those *DSM* codes *certainly* don't know whether they're real.

A 2012 *Canadian Medical Association Journal* study showed that within a grade, schoolkids were far more likely to be diagnosed with ADHD merely *by being younger*. As the study put it: "Boys who were born in December were 30% more likely to have a diagnosis of ADHD than boys born in January, and girls born in December were 70% more likely to have a diagnosis of ADHD than girls born in January." As family therapist Joan Lipuscek observes, immaturity was being misdiagnosed as ADHD.

If childhood bipolar is arguably not a real diagnosis, what's to say that ADHD is the right rubric for what may be multiple ailments or even no ailment at all? What is the reason for sticking with current *DSM* categories rather than the alternative paradigms proposed by *DSM* opponents like Edward Shorter? Are mania, depression, and bipolar three separate things, or all aspects of one overarching mood disorder, which Shorter terms Kraepelin's disease, or "melancholic syndrome"? How do we even adjudicate such issues, in the absence of decisive neurobiological evidence?

Frances advocated a conservative approach, believing that changes to the *DSM* would exacerbate diagnostic ambiguity by inflating the number of diagnoses without helping to ground them more firmly in neurobiology. Frances's advice, then, is to diagnose less and treat less, using the *DSM* provisionally: "The right goal for *DSM-5* would have been diagnostic restraint and deflation, not a further unwarranted expansion of diagnosis and treatment." In this, however, he was at war with the *DSM-5* leaders who both found the existing categories inadequate *and* had the ambition to set things right. And this is where computers enter the *DSM* story.

Computers were a driving force behind this latest, troubled revision of the *DSM*. From the *DSM-5*'s conception in 1999, the American Psychiatric Association's *DSM-5* architects, led by chairman Darrel Regier, had grand ambitions to adopt new *dimensional* measures of diagnosis instead of categorical ones. Diagnoses were to be graded on a sliding scale (How depressed are you? How severe is your schizophrenia?) rather than by yes-or-no criteria. The *DSM-5* team, including Regier and Alan Schatzberg, rebutted an objection by Frances by appealing to a 2009 article co-authored by Regier:

> As documented in the recent *American Journal of Psychiatry* (*AJP*) article, the use of dimensional assessments to reconceptualize psychopathology represents the most practical and evidence-based way of moving our field forward.

That 2009 article shares two of the same authors and promotes dimensional measures without providing any evidence that they're more successful:

> Our immediate task is to set a framework for an evolution of our diagnostic system that can advance our clinical practice and facilitate ongoing testing of the diagnostic criteria that are intended to be scientific hypotheses, rather than inerrant Biblical scripture. The single most important precondition for moving forward to improve the clinical and scientific utility of *DSM-V* will be the incorporation of simple dimensional measures for assessing syndromes within broad diagnostic categories and supraordinate dimensions that cross current diagnostic boundaries.

In plain English, this passage claims dimensional measures will be a crucial improvement to the *DSM*.* The justification for dimensional measures tended to center on a 2006 paper coauthored by another *DSM-5* architect, Helena Kraemer, a biostatistician in Stanford's department of psychiatry whose background is in statistics rather than medicine. As the dimensional measures mostly did not make it into the *DSM-5*, I won't relitigate this dispute. What interests me more is that the implicit rationale for this shift to dimensional measures was increased quantitative specificity. Kraemer sought to make the diagnostic process as predictable and standardized as possible. To her, dimensional measures offer

* The numbing verbiage of the leaders of the *DSM-5* contrasts to the considerably more forthright writings of their predecessors. A similar apologia by Regier's colleague William Carpenter similarly masks a paucity of content in vague obfuscation: "The challenge for *DSM-V* is whether an approach to domains of pathology can supplement syndrome diagnostic categories. This does introduce a new paradigm—domains of pathology—and the scientific field is moving rapidly to deconstruct heterogeneous syndromes." The use of the word "deconstruct," signaling a heady revisionary mind-set, is almost always a red flag—the presence of "paradigm" doubly so.

> a new set of advantages such as greater statistical power, improved predictive validity, more focused treatments, and new opportunities for genetic and other etiological research.

These "advantages" share several characteristics. All of them restrict the individual initiative of a physician. All appeal to harder science than psychiatry has offered to date. And *all of them require computers.* The statistical analyses and research opportunities Kraemer cites all require *the collection of data* in order to assess the reliability and utility of treatments. The goal is to increase the granularity of wide-scale (and narrower-scale) data collection, so that uncontrolled field studies can be performed to a far greater degree of precision, supplementing controlled studies. Kraemer bemoans the "forfeit of clinical information" caused by the existing *DSM* categories:

> Perhaps the greatest disadvantage of a categorical system is the limitations, both clinical and statistical, imposed by the forfeit of clinical information inherent in labeling patients based solely on whether their signs and symptoms collectively rise above a defined threshold.

To get a grasp on an individual patient, one may need to read a physician's notes or even speak to him or her. But for Kraemer, "clinical information" is *only* what can be represented by the *DSM* categories. It is purely quantitative. Kraemer focuses on the transcription of patient data *into* a computer, even offering a vision of clinical assessment by computer:

> Consistency in the collection of clinical and epidemiological data would be enhanced if structured interviews and/or questionnaires were developed and offered as part of the *DSM-V*. Patient self-administered questionnaires could be developed to gather relevant symptom data. Such efforts would offer even greater benefit if users had the choice of paper or computer administration. There is considerable evidence that responses to computerized interviews are more candid than face-to-face responses.

Kraemer's goal, then, is the quantification of the diagnostic process itself so that diagnosis could be possible even in the absence of a human physician. Diagnosis becomes another version of a standardized test one takes in school. The benefits are uniformity of data being collected and removal of potential bias in the physician. Unfortunately, this isn't possible. Kraemer speaks of "a direct, dimensional reflection of the categorical definition that could be used for genetic and other analyses to increase statistical power." But if the categories are bunk, there's no way that chopping them into spectra is going to make them better. The categories will only become more arbitrary and detached from the everyday realities that spawned them. Psychotherapist Gary Greenberg's account of the dismal and sometimes comical results of the *DSM-5* task forces, where the members often could not agree on what it was they were trying to accomplish, shows this detachment stymieing the participants repeatedly.

The *DSM-5* project did not break down just because it was a cultural construct. We *need* cultural constructs to organize our lives, and we inevitably create them in their absence. Rather, the *DSM-5* couldn't satisfy several competing interests at the same time. The *DSM* today is used for physician diagnosis, actuarial insurance practices, longitudinal research studies, drug regulation, and more. Robert Spitzer and his colleagues architected the *DSM-III* with one primary purpose in mind: standardizing diagnoses across physicians. Yet because its classifications are so provisional, the *DSM*'s comparative success at *that* goal doesn't mean that it was or is suited for the many other purposes to which it is being put. The differing interests and rationales make consensus impossible. There will be a tug-of-war between people trying to pull these provisional categories in different directions to suit differing agendas, and no way to claim that one direction is "right," because there is no agreed-upon truth to measure a classification against. The *DSM* has passed its expiration date not because it is intrinsically nefarious, but because the context in which it was created no longer exists, having been replaced with newer models of psychiatry, and it is failing to meet the requirements currently being put upon it.

The National Institute of Mental Health is abandoning the *DSM* as a diagnostic taxonomy in favor of the less politically loaded and less specific categories of psychiatric disorders listed in the *International*

Classification of Diseases—a classification owned and managed by the United Nations' World Health Organization rather than the American Psychiatric Association. In 2013, NIMH head Michael Insel wrote of the decision:

> While *DSM* has been described as a "Bible" for the field, it is, at best, a dictionary, creating a set of labels and defining each. The strength of each of the editions of *DSM* has been "reliability"—each edition has ensured that clinicians use the same terms in the same ways. The weakness is its lack of validity. Unlike our definitions of ischemic heart disease, lymphoma, or AIDS, the *DSM* diagnoses are based on a consensus about clusters of clinical symptoms, not any objective laboratory measure. In the rest of medicine, this would be equivalent to creating diagnostic systems based on the nature of chest pain or the quality of fever. . . . Patients with mental disorders deserve better.
>
> The diagnostic system has to be based on the emerging research data, not on the current symptom-based categories.

To Gary Greenberg, Insel was even more frank: "There's no reality. These [disorders] are just constructs. There's no reality to schizophrenia or depression. . . . Whatever we've been doing for five decades, it ain't working." And yet even while recognizing this, Insel did not toss out taxonomies entirely, but turned to what he thought of as a better provisional taxonomy, the *ICD*.

Without scientific consensus on the nature of mental illness, we cannot do without folk taxonomies like the *DSM* or the *ICD*. But we must recognize, as Frances says, both their limitations and their dangers. Given the increasing quantification of every aspect of daily life, we will suffer similar difficulties considerably more often. Even if the *DSM* is recognized as obsolete, terabytes of medical records containing *DSM-III*, *-IV*, and *-5* codes remain, ready for research and actuarial purposes. Computers make it so that any taxonomy like the *DSM* will hang on far longer than we may wish it to. Computers reify existing taxonomies, making them harder to displace, while simultaneously making them appear more adequate—and accurate—than they actually are.

6

GAMES COMPUTERS PLAY

Dungeons and Dice

> What is it that you see when you see? You see an object as a key, a man in a car as a passenger, some sheets of paper as a book. It is the word 'as' that must be mathematically formalized, on a par with the connectives "and," "or," "implies," and "not." . . . Until you do that, you will not get very far with your AI problem.
>
> —STANISLAW ULAM

IN THE EYES of the *DSM,* we are a collection of diagnoses. In the eyes of Myers-Briggs, we are one of sixteen personality types. In the eyes of another human, or indeed ourselves, we are a countless number of things that, when taken together, approximate our essence. To be seen is to be seen as something. Computers are incapable of seeing us as people do, but they are very good at seeing us as numbers.

The story of the *DSM* is the story of fitting human pathology into a quantitative taxonomy. It is a messy business, because the labels never match up exactly, and it makes for a gross simplification. But what if we were to begin from the other side? What if, instead of trying to regiment human complexity into similar forms, we try to build out human complexity from computer-friendly representations?

There was one other famous system for the quantification of human personality and behavior that I became familiar with as a child, which was Dungeons & Dragons (or Advanced Dungeons & Dragons, as it was known in its standard incarnation back then), created by Gary Gygax out of the residue of dozens of less successful antecedents. The mainstreaming of geek culture has turned both computers and role-playing (whether on computer, tabletop, or "live-action" LARPing) into major social phenomena, such that it's very difficult to tease apart the development of each. Systems like D&D appealed to the analytic mind-set of many engineers. What they provide, after all, are the computational algorithms for an entire fantasy world.*

D&D and other role-playing games (RPGs) share a common core. Half a dozen or so people each adopt an ongoing alter ego whom they play over multiple game sessions. The Dungeon Master (DM), who referees and drives the game, provides some kind of scenario for these characters to explore jointly. The characters negotiate the traps and monsters laid out by the DM with their individual skills, and the characters develop and become more powerful through their adventures. Characters can also die. A great deal of responsibility is placed on the DM, who has to challenge the players without frustrating them and create suspense without being cruel.† Unlike sports games, where the referee is only there to apply rules, DMs have a great deal of latitude to bend or even break rules at their discretion—such as, for example, saving a character when death would disrupt play too much—or alienate a participant.

An explicit set of numerical mechanics, manipulated through simple arithmetic, provides the glue between D&D's structuring elements. The backbone of D&D's mechanics is its class and stat system. A player picks a class (fighter, thief, wizard, etc.) to determine a character's basic

* I did think it was delightful that D&D made use of Platonic solids for its dice rolls: the D4 tetrahedron, the D6 cube, the D8 octahedron, the D12 dodecahedron, and the D20 icosahedron. I recall the D12 being the wallflower, especially given the incredibly frequent use of the D10, a non-Platonic ten-sided die with kite-shaped sides. (It is far more ungainly than the Platonic solids.) Two D10s were handy for making rolls against 100-entry tables, of which D&D had *many*.

† Or not. Gygax's infamous 1975 adventure *Tomb of Horrors* was intentionally designed to kill off characters who had gotten too big for their britches—including that of Gygax's own son Ernie.

skill set (punching, casting spells, pickpocketing), then rolls dice to determine how good the character is at each skill. D&D contains a core set of six character statistics. There have been hundreds of variations on the basic D&D template, but most don't stray too far from these original six, which divide into three physical and three mental (though since magic exists, mental powers can be pretty physical too):

Strength: How hard do you bash things?
Dexterity: How fast do you bash or avoid being bashed?
Constitution: How much bashing can you take?
Intelligence: How book-smart are you (i.e., how hard do your spells bash, and how quickly do you learn new ways to bash things)?
Wisdom: How street-smart are you (i.e. how's your bashing-avoidance know-how)?
Charisma: How good are you at convincing people not to bash you?

The algorithms of D&D are simple: a "stat check" consists of rolling a die and not rolling over the character's current skill level. A character's ability to learn woodworking from a woodworking book would require a 1d20 roll against intelligence: if the roll of a 20-sided die results in a number lower than the character's intelligence value, the character learned woodworking. But if my intelligence was 12 and I rolled a 16, my character wasn't intelligent enough to learn woodworking. Need to cross a rickety bridge? Dexterity check. Decipher ancient runes? Intelligence check. Play the alpenhorn? Constitution check.

Everything in D&D is quantified. Armor has a number indicating how tough it is, a weapon has a number for how much damage it does, and every person or monster in the game has its own set of numerical stats. There are numbers for how good characters are with particular weapons, particular tools, particular skills, particular jobs. Everything functions algorithmically by comparing dice rolls against one stat or another, though players have only partial knowledge compared to the Dungeon Master, who tracks the whole story and keeps certain dice rolls secret. I might know that my Dual-Tipped +3 Enchanted Quarterstaff of Pusillanimity did 2d8+3 damage against the Pulchritudinous Troll, and I could calculate that after rolling a 5 and a 2 with an 8-sided die that I'd done a raw 10 points of damage (5+2+3), but the Dungeon Master might tell me only that the Pulchritudinous Troll was looking a bit more haggard, leaving me uncertain about how much more damage I needed to do while the troll bashed me for 1d20+5 (6 to 25) points of damage every turn. After killing the troll, my character gains experience points and perhaps a level of experience, at which point my character could also earn new abilities and skills. The gradual leveling of a character provides an ongoing sense of progress and mastery that permits greater challenges to be met (tougher monsters, mostly).

For me these challenges weren't so thrilling. In computerized versions, where progress and acquisition were greatly accelerated compared to the human tabletop game, I could appreciate the novelty of finding an Infinity+1 Sword or acquiring the powerful yet impractical Armageddon spell, but much of the actual gameplay was tedious. Classic series from the eighties like *The Bard's Tale, Ultima, Wizardry,* and *Might and Magic,* which were all based on D&D-like models, consisted of endless battles in twisty dungeon mazes to "level up" your characters so they could survive being bashed by ever more powerful mon-

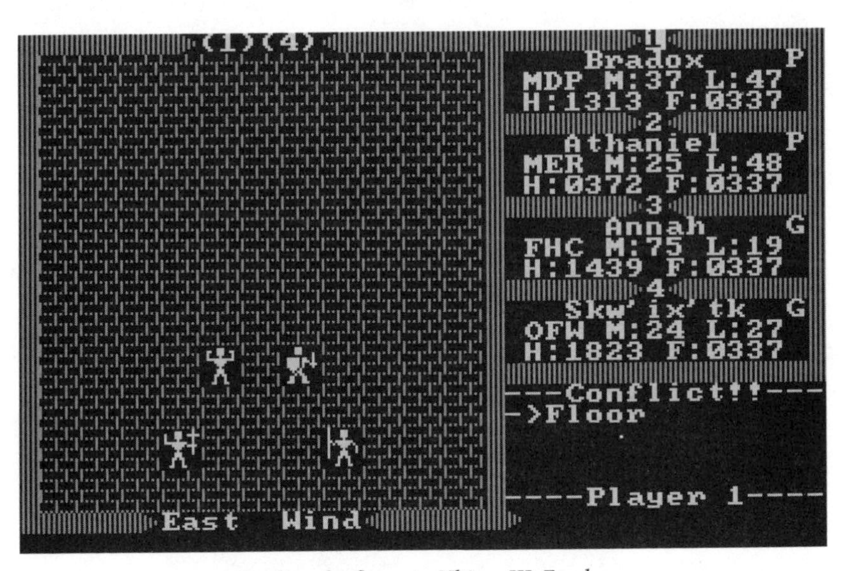

Battling the floors in *Ultima III: Exodus*

sters. This process is known as "grinding," and I never cared for it. The tedium outweighed the strategy. The details I remembered most were the incidental ones, like the bizarre "floors" from *Ultima III,* enemies you couldn't see because they looked just like the floor. Invisible moving monster floors: that's a great touch.

The diabolically difficult mazes of *Wizardry IV* were more interesting for me to contemplate than they were to play. (This was a game that was so difficult most people were unable to figure out how to get out of the first room.) I marveled at people who enjoyed agonizing trial and error, frustration, and endless character death as player retention strategies. I didn't enjoy the gameplay, but I was fascinated by the games' simplified representations of reality.

D&D's stats are the data structures of D&D. In the 1970s, when D&D was strictly played on paper, the game was already well suited to computers. D&D's metrics of character and world development, represented *numerically* (for stats) and *geometrically* (in dungeon layout and combat), provided a combination of simplicity and precision that made the adaptation of tabletop role-playing to computer gaming inevitable. No other fictional genre, not even science fiction, was as tailor-made for algorithmic computation as the fantasy world that D&D elaborated. D&D relaxed fiction's burden of storytelling and character psychology in favor of coarse but *exact* mechanics, simplifying the world into a set of representations that a computer could manage far better than, for example, the mystifying world of mental illness. Yet both D&D and *DSM* embrace *approximate quantified fictions.* They are both provisional, extensible, and revisable folk taxonomies.*

D&D's extensive auxiliary components quantify its world, borrowing lore from mythologies and religions of every stripe, resulting

* The tropes of D&D, its Tolkienized antecedents, and its many offshoots have spread through computer games far and wide, while those of almost no other genre, whether highbrow literature or lowbrow pulp mystery, have. The high-fantasy source material of Dungeons & Dragons, based in the pulps of Robert E. Howard's Conan, L. Sprague de Camp and Fletcher Pratt's whimsical fantasies, and above all J. R. R. Tolkien's medieval fantasy cosmos (where cultural detail was more important than plot or character) proved uniquely suited to gaming and then to computation. The combination of physical combat, arbitrary and easily contrived magic, imaginary landscapes, and elaborate taxonomies of trait, skill, and race all contributed to making high fantasy the default interactive world of the computer. By focusing on environment and mechanics over human relationships and politics, Tolkien et al. unwittingly set the stage for computer-facilitated worldbuilding. In no other genre is it so easy to see how the rolling of dice—the generation of random numbers—could become *the* core mechanic of gaming.

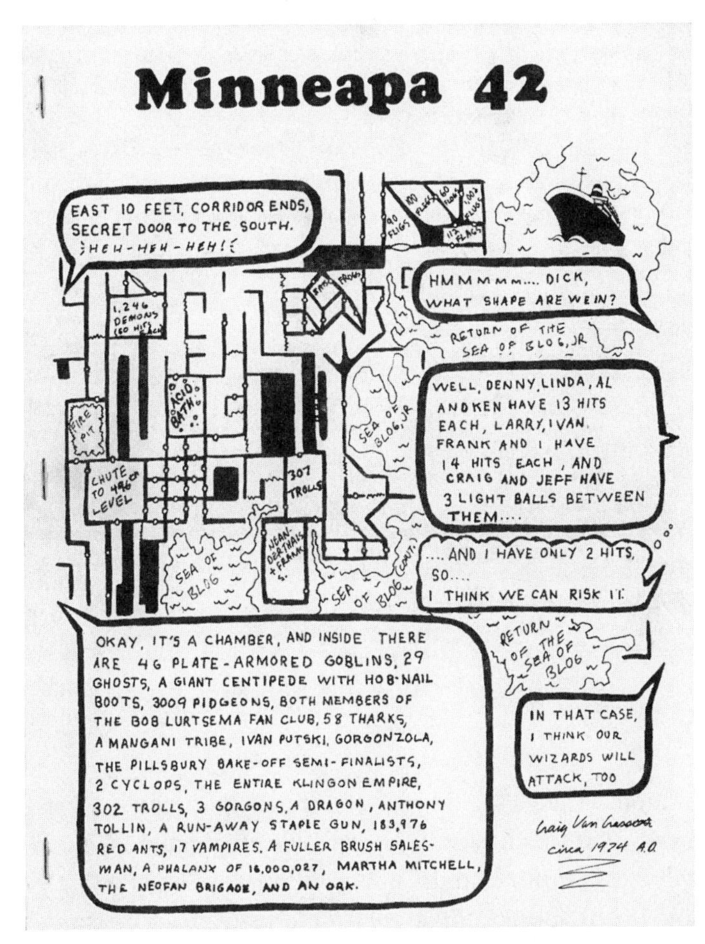

April 1974: The *Minneapa* fanzine's parody of D&D's system of approximate quantifications

in a perverse syncretic assemblage.* Gygax was far more a synthesist than an original thinker, and his genius lay in his doggedness at creating kitchen-sink worlds capable of engaging role-players of every stripe, worlds simultaneously open-ended and simple enough to allow

* Game historian Jon Peterson briefly alludes to theosophy as a possible precursor to such syncretism. Peterson gives great support to the notion that Gygax, in presiding over the yoking of Tolkienesque fantasy mythology to military wargaming mechanics, is responsible for initiating a cultural empire akin to Walt Disney's (who had analogously yoked children's entertainment and fairy tales to animation), though Gygax's comparative lack of ruthlessness prevented him from retaining control over its development.

Harlot encounters can be with brazen strumpets or haughty courtesans, thus making it difficult for the party to distinguish each encounter for what it is. (In fact, the encounter could be with a dancer only prostituting herself as it pleases her, an elderly madam, or even a pimp.) In addition to the offering of the usual fare, the harlot is 30% likely to know valuable information, 15% likely to make something up in order to gain a reward, and 20% likely to be, or work with, a thief. You may find it useful to use the sub-table below to see which sort of harlot encounter takes place:

01–10	Slovenly trull	76–85	Expensive doxy
11–25	Brazen strumpet	86–90	Haughty courtesan
26–35	Cheap trollop	91–92	Aged madam
36–50	Typical streetwalker	93–94	Wealthy procuress
51–65	Saucy tart	95–98	Sly pimp
66–75	Wanton wench	99–00	Rich panderer

An expensive doxy will resemble a gentlewoman, a haughty courtesan a noblewoman, the other harlots might be mistaken for goodwives, and so forth.

D&D's Harlot encounter table

for variation.* There's the simple yet uniquely peculiar two-axis alignment system, in which each character is a combination of Good/Neutral/Evil and Lawful/Neutral/Chaotic. Similarly, Gygax had a passion for using the basic algorithms of D&D to regulate every aspect of the created gameworld. Though boiling down every player interaction to numerical dice rolls was hardly *necessary* for play, the ubiquitous dice rolls provided a centralized gameplay mechanic without restricting how that mechanic could be used. Gygax, for his part, was indefatigable and enthusiastic in constructing these suggestive, loose-fitting analogues to the world.

The drollest example I know of Gygax's omnivorous yet haphazard rage for classification and quantification is the "Harlot encounter" table from the first edition, written by Gygax himself. It is a table that

* The downsides of Gygax's more-is-more approach are evident in his little-known three-dimensional chess variant *Dragonchess,* which plays out on three levels, each of which is an oversized chessboard, and gives each player forty-two pieces instead of the usual sixteen, all with complicated and seemingly arbitrary rules for moving. All of Gygax's creations possess too much of a muchness.

indexes dice rolls from 1 to 100 (using two ten-sided dice), and it demonstrates conventional wisdom and cultural norms being put through the wringer of quantification.

The ordering of the harlot taxonomy goes by what could loosely be termed the "classiness" of the harlot. Number divisions usually have an implicit or explicit rank to them. Rolls from 01 to 90 result in harlots of increasing price. Rolls from 91 to 00 produce pimps of four varieties, though for some reason a "sly pimp" is twice as likely to occur as any of the other three types. With the exception of the haughty courtesan, who has only a 5 percent chance to show up, other types of harlots occur with either 10 or 15 percent probability. From this we see that there are nearly as many sly pimps in the world of D&D as there are haughty courtesans. I'm not sure how many Dungeon Masters would be able to assess and depict a meaningful difference between a saucy tart and a wanton wench, which is likely one of the reasons why this table disappeared from subsequent editions of the game. There is, indeed, little reason for this table to exist beyond a sheer mania for classification itself.* But what makes a "haughty courtesan" higher-class than an "expensive doxy"? How does a DM distinguish between a "slovenly troll" and a "brazen strumpet"? As with the *DSM,* the classifications are in large part arbitrary.

But as the harlot table suggests, the quantifications are also soaked in conventional wisdom. Just as homosexuality entered and then exited the folk taxonomy of the *DSM*'s pathologies, both fashionable and unfashionable ideas fell into D&D as "objective" descriptions of reality. Nowhere is this more apparent than in the treatment of gender. Historically, wargaming and tabletop gaming were overwhelmingly the province of men, a product of the male hobbyist origins of the games. Women entered the workforce and academia—even in computer science—decades before making similar levels of incursions into male recreational activities. As the hobbyist and academic worlds of computers merged after the dawn of the home computer in the 1980s, computer science became more gender-imbalanced, with the number

* Gygax's own torn loyalty between quantified constraints and creative freedom is evident in two infamous quotes, "A DM only rolls the dice because of the noise they make," and "The secret we should never let the gamemasters know is that they don't need any rules." Peterson also quotes Chuck Ulrich's creation of the officious, bull-headed race of "gygacks," who "will insist upon everything being done their way, although they will insist that they favor individuality and diversity."

of undergraduate degrees awarded to women nearly dropping in half between 1986 and 1990. The recreationalization of the computer was likely a significant cause. The world of Dungeons & Dragons was very much the product of an overwhelmingly male mind-set that catered overwhelmingly to men. Yet D&D's flexibility allowed it to evolve to accommodate female players, mostly by eliding the distinctions between male and female characters. Longtime D&D writer/fan Lenard Lakofka's 1974 system for developing female characters in D&D suggested that female characters be limited in strength to 14, while males were at 18, along with several other recommendations that infuriated many female and male players. Lakofka, alongside Gygax and *Dragon* magazine's publisher, were hanged in effigy in a cartoon in the following issue. When a system is such an inexact approximation of reality, *any* such distinction becomes laden with value judgment and prejudice. As a result, sex distinctions were collapsed within D&D, and only fantasy ones (between elves and humans, for example) were maintained. Ironically, as sex differences dissolved, cultural and physiological differences between races (half-orc, elf, dwarf) grew increasingly broad and extreme.

By far the most fascinating D&D creation I know is the bizarre feverdream *Planescape: Torment,* originally released in 1999, which tosses out most of the stat-building and a good chunk of the combat in favor of a twisted and often deranged morality tale of amnesia, immortality, remorse, and psychosis. *Paradise Lost* is thrown in as an incidental secondary plot. I tore through *Torment* with a combination of respect and disbelief, amazed at the uncontrolled and unrefined passion that made it difficult to anticipate what craziness it would throw at me next. (There are over a million words of text in the game, and they range very far afield.) There were dream sequences, fake religious cults, magical tattoos, sentient rat kings, doppelgangers, robot mazes. The stats take a backseat, as by the end of the game the main character has "leveled up" so drastically he has godlike powers, a total betrayal of the slow-grind progression that typifies role-playing games like D&D. *Torment* has a well-deserved cult following. Yet it (as well as similarly bizarre creations like *Vampire: The Masquerade–Bloodlines* and *Mask of the Betrayer*) would not have come into existence except as a response to the far duller rudiments of most RPGs (*The Bard's Tale, Diablo, War-*

craft, Skyrim). We need our arbitrary frameworks, even the mediocre ones, for creativity to soar.

Planescape: Torment gains its resonance from choosing the *right* threads to follow in Gygax's loose tapestry, opting for surrealism and subversion at every juncture and ignoring the less interesting bits. The unsettling psychological derangements within the game owe little to the peculiar and antiquated "Types of Insanity" table from Gygax's original rules, listing the conditions that could be triggered in a character due to "mental attack, curse, or whatever."

TYPES OF INSANITY

1. dipsomania
2. kleptomania
3. schizoid
4. pathological liar
5. monomania
6. dementia praecox
7. melancholia
8. megalomania
9. delusional insanity
10. schizophrenia
11. mania
12. lunacy
13. paranoia
14. manic-depressive
15. hallucinatory insanity
16. sado-masochism
17. homicidal mania
18. hebephrenia
19. suicidal mania
20. catatonia

Gygax's table does not derive from a single source. Coming up to a round twenty diagnoses, Gygax must have either combined or trimmed sources to get to that convenient number. In some respects, it is similar to the then-current *DSM-II*, but it also includes illnesses like dementia praecox and melancholia that had been eliminated in the *DSM* but look back to psychiatrist Emil Kraepelin's original taxonomy of psychoses formulated at the end of the nineteenth century. Gygax doesn't acknowledge, realize, and/or care that dementia praecox and schizophrenia were generally considered identical, or that catatonia and hebephrenia were subcategories of schizophrenia (and dementia praecox). Not that Gygax's definitions bear much relation to the clinical ones: Gygax defines dementia praecox as ennui, while schizophrenia is based on the then-common conflation with split personalities:

> **Schizophrenia:** This form of insanity has the well-known "split personality" trait. From 1 to 4 separate and distinct personalities can exist in the afflicted—base the number upon the severity of the insanity. Likewise, the difference from one personality to the next should reflect the severity of the affliction. Each "new" personality will be different in alignment, goals, and preferences. (A very severe case might have a different class also but without coincidental possession, the new personality emerging will not have the actual abilities he or she may think that he or she possesses.) The onset of schizophrenia is random, 1 in 6 per day, with a like chance of a new (or return to the old) personality emerging. However, whenever a stress situation—decision, attack, etc.—arises, the 1 in 6 chance of schizophrenia striking must be checked every round in which the stress continues.

"Naturally," Gygax writes, "these forms of insanity are not clinically correct." It seems unlikely that Gygax chose the names in the insanity table for any reason beyond their mythic and exotic sound. In the game, mental disorders are regimented to be more predictable than in real life: manic "lunacy" is literally triggered by the moon's phase, while sadomasochism causes the player to alternate between sadistic *and* masochistic periods every three days. The drearier names of the *DSM-II* are absent from the list: no one in the world of D&D suffers from obsessive-compulsive disorder, hypochondria, or even anxiety.

Longtime D&Der Bob Kruger describes the appeal and practice of the game as follows:

> Essential D&D resonates with our evolved cognitive biases, and explores archetypal symbols. It takes practice to do it well, but doing it well is less a function of adhering to the rules than mastering the art of appearing to faithfully apply them.

Kruger strikingly echoes the rationales for provisional psychological taxonomies like the *DSM*. A folk taxonomy based in human cognitive predispositions becomes an explicitly delineated rule system. Yet the system is full of arbitrary kludges, and the rules are there not for *strict* obedience but for loose guidance—just as most physicians treat the *DSM*. And in both cases, transferring the ruleset to computers removes the ability to slacken the rules ad hoc. Computers reify the taxonomy in the most literal sense: computers *make the taxonomy more real*. The taxonomy becomes more than a set of labels. It becomes a systematic organization of existing things: an *ontology*.

In other words, the stats of D&D aren't just a tool for playing a game. They function as the interface to make the gameplay more deeply meaningful to a player. The stats assist in *representing* that player's character, who in turn *plays* the role of the character the stats describe. Jon Peterson describes the core appeal of role-playing games as offering players the ability "to discover a persona worth embodying" via "modeling imaginary people." When labels and numbers are piled atop one another in service of describing *people*, whether in MBTI, MMPI, OCEAN, *DSM*, D&D, or any other system, their folk ontologies offer us personae to play.*

The limits of D&D were, in large part, the limits of the Dungeon Master. A DM could keep track of only so much data about the players and the creatures they encountered. So while D&D expanded, the number of factors in play at any one time didn't. This acted as a hard limit on just how complex the underlying mechanics of a game could be. The human work required to bring D&D closer to the complex-

* Echoing *Rameau's Nephew*, Erving Goffman's *The Presentation of Self in Everyday Life* posits that social environments depend on preexisting specifications of social roles. "When an individual or performer plays the same part to the same audience on different occasions, a social relationship is likely to arise. Defining social role as the enactment of rights and duties attached to a given status, we can say that a social role will involve one or more parts and that each of these different parts may be presented by the performer on a series of occasions to the same kinds of audience or to an audience of the same persons."

ity level of "reality" would be not just time-consuming, but hopelessly tedious. Computers, however, thrive on tedium.

Deterrence and Détente

> Now learn how very frail an image is.
>
> —LUCRETIUS

As originally conceived, the mechanics of D&D-like tabletop role-playing games had to be *transparent* to players: the numbers and arithmetic had to be obvious and in plain sight, so that what happened in the game *made sense*. D&D was extensive yet mechanically simple. The danger of having overly complex and rigid rules is twofold. First, they make the Dungeon Master's job a nightmare, stifling creativity while demanding meticulous analysis of player actions. Second, they make it harder for players to understand why their actions are having the effects they are having, because the calculations are opaque. If a player gets whomped by an orc, the player should have been able to assess the danger of the situation ahead of time, even if the exact numbers are unknown. D&D keeps cause and effect clear.

But with a computer acting as the Dungeon Master, entire layers of simulated reality can be opaque to the players, just as in real life we go about unaware of the semiautonomous workings of our body's systems and the chaotic processes of the weather. With the management of a computer, the behavior and appearance of a harlot no longer need to be obtained from a single number between 1 and 100, but can be determined through a more complicated model of personality and the social world, incorporating economics and genealogy.* It need not be a harlot; it could just as easily be a monster, or a corporation, or a nation.

My introduction to these more complex models came through a classic simulation game of the 1980s, Chris Crawford's Cold War geopolitical simulator *Balance of Power*. Originally published in 1985, it

* Of course, the creation of such a deep world requires great programming effort, and I'm not aware of any computer RPGs that have bothered to model the socioeconomic reality of prostitution.

was critically acclaimed, even garnering a review from Jimmy Carter's deputy national security advisor, David L. Aaron, in the *New York Times*. Aaron took note of the game's simplifications, but praised it for being both compelling and educational. In taking on the role of either the United States or the USSR, a player tries to win over some of the eighty non-superpowers either directly or indirectly. If you meddle too much in your opponent's sphere of influence, they would object, at which point you could either back down (at the cost of "prestige," which was more or less your score in the game—your ability to influence world affairs), or you could . . . escalate. The computer in turn would either back down or escalate, and DEFCON 3 would turn into DEFCON 2, which would turn into DEFCON 1. I remember the screen that greeted me when DEFCON 1 was reached:

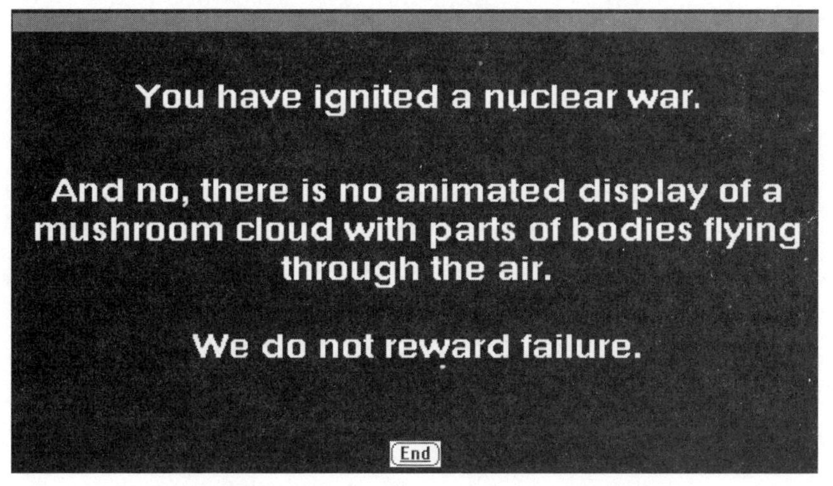

The result of a game of nuclear chicken in *Balance of Power*

Anyone who played the game saw this screen a lot. In my first game playing as the United States, I sent troops into Iran to fight for the rebels, oblivious to more or less every geopolitical consideration, and saw the Soviets immediately object. Boom. Then the Soviets sent troops into a civil war in *Brazil*, which I thought absurd. They didn't back down. Boom. Aaron had a similar experience: "A dozen tries later, I was still destroying the earth with depressing consistency. . . . The more I played *Balance of Power*, the more my self-destruction stemmed from an unwillingness to back down in a crisis." Every conflict was a Cuban

Missile Crisis, and the chief lesson of the game, inasmuch as there was one, was that in a nuclear world, one should pick one's battles *extremely* carefully.

My greater fascination, though, was with the book Crawford wrote that explained the math the computer used to determine whether to push its luck. The computer player's logic was based on a calculation of *Outrage* over a particular player action:

> Suppose that a crisis erupts over some truly insignificant action, such as economic aid to Nigeria. Players often find to their dismay that the computer will escalate right up to DefCon 1 in such situations. Why, they complain, would the computer destroy the world over such a trivial issue?
>
> The answer is, because *you* would destroy the world over such a trivial issue. The computer analyzes the conflict and finds that its *Outrage* over the issue is small, such as 22. It finds, however, that the human pleasure over the issue is even less, such as –18. When it adds the two numbers together, it gets a +4 result and concludes that it is justified in taking a firm stand. If the human can ask, Why would you destroy the world over a measly 22-point crisis? the computer is even more justified in asking, Why would you escalate to DefCon 2 over something that was worth only 18 points to you? It takes two to make a fight.

I was interested not only in the algorithm, but also in how Crawford had chosen to calculate outrage based on six other variables:

$$\text{Outrage} = \frac{\text{Hurt} * (\text{Obligation} + \text{DipAff}) * \text{DontMess} * \text{Adventurousness} * \text{Prestige Value}}{128}$$

Slight elements of randomness were used to make the game nondeterministic, but the model was tight. It was the product of Crawford's mind, but it was a mind that had read George Kennan, Henry Kissinger, and others, and had then engineered a playable model of the dynamics they described. The game felt mechanical, yet it showed that such mechanical models could do more than approximate how hard a barbarian could bash an orc—they could also present abstract psychological and political concepts. The geopolitical environment that evolved over a game of *Balance of Power* felt nearly organic. Its limitations were primarily those of the computers of its time. There was only so much space on a floppy disk and only so much processing power on the computers for which *Balance of Power* were written, and the complexity of its geopolitics was limited by the requirement that the game spend seconds, not minutes, calculating the effects of the player's actions each turn.

The Quantified Dwarf

> Our empirical language can only be understood as an incoherent and fragmentary schema of an ideally coherent language.
>
> —WILFRID SELLARS

The computational limitations that *Balance of Power* faced are no more. Computers are over ten thousand times faster now than they were in 1985. Data-processing capacities are not quite unlimited, but when it comes to simulations, computers now offer the ability to construct worlds that do not reduce to a handful of equations, but are complex ecosystems in and of themselves. The average computer role-playing game today is far richer *algorithmically* than *Balance of Power* or D&D. "Physics engines" provide realistic modeling of three-dimensional motion. Game action can take place in real time rather than requiring sequential, turn-by-turn choices. The drive for verisimilitude and precision, combined with the sheer computational power provided by PCs, has allowed for the creation of virtual worlds beyond what Gygax ever could have crafted by hand. These worlds contain quantitative systems so complex that their interactions can lead to counterintuitive and unpredictable results—even when the individual systems themselves have been coded logically and intuitively. Such worlds can begin to compete with our own—and imitate it.

The mania for quantification reaches its peak in the game *Dwarf For-*

Tim Denee's schematic of his *Dwarf Fortress* "Oilfurnace"

tress, which, despite the fantastic name, is one of the most meticulously literal-minded simulations around. The premise is simple: the player is the overseer of a band of a few dozen drunken dwarves, who attempt to build and secure a safe homestead inside a mountain, navigating the dangers of natural disasters, hostile wildlife, invaders, and their own sheer irresponsibility and incompetence. The player issues orders, but the dwarves themselves are autonomous entities who act based on hidden psychological modeling. They may obey, disobey, lose hope, cheer up, eat, drink, fight with one another, reproduce, make art, panic, and (rather frequently) die of their own volition. You lose when all your dwarves die; the game has no success conditions other than continuing to survive. The brainchild of auteur Tarn Adams, who estimates the game will take another twenty years to complete, *Dwarf Fortress* makes no concessions to commerciality or accessibility and subsists through the support of a group of diehard fans, who see the game as the ne plus ultra of strategic simulations.* Its level of complexity would be impossible on the tabletop, but a computer can model a dwarf's body, determine whether an attack has bruised the dwarf's arm, collarbone, or kidney, and factor that into the ongoing game state, even if the player is not made aware of these conditions.

The design goal of *Dwarf Fortress* is to create a fantasy world that behaves *as the real world does*. The game has its own internal models of physics, biology, agriculture, weather, bodies, creature motivations, and even chemistry, so that when a surprising event occurs, it will nonetheless *make sense* in that it relates to something a player would expect from that situation in real life.

PAUSED
Dwarf Fortress
Idlers: 7
a: View Announcements
b: Building r: Reports
c: View Civilizations
d: Designations o: Set Orders
k: Look Around j: Job List
m: Military s: Squads
N: Points/Routes/Notes
w: Define Burrows
p: Stockpiles i: Zones
q: Set Building Tasks/Prefs
R: View Rooms/Buildings
t: View Items in Buildings
v: View Units h: Hot Keys
n: Nobles and Administrators
z: Status
Tab: Move this menu/map
?: Help Escape: Options
;: Movies D: Depot Access
Space: Resume .: One-Step

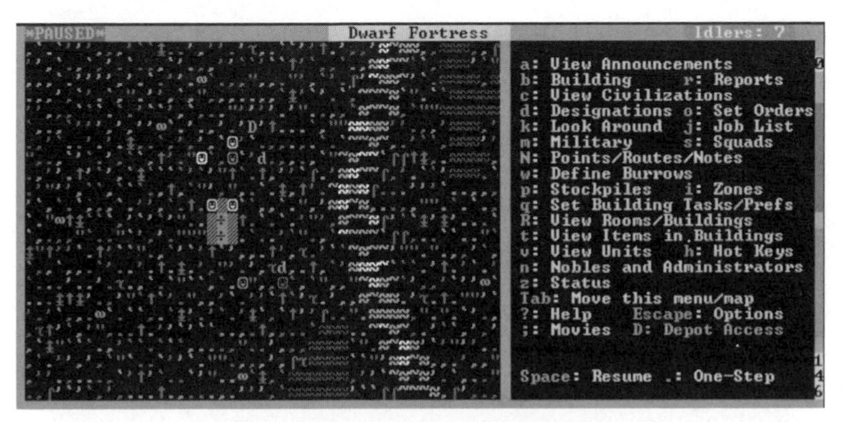

The standard interface of *Dwarf Fortress*.
The seven happy faces are dwarves.

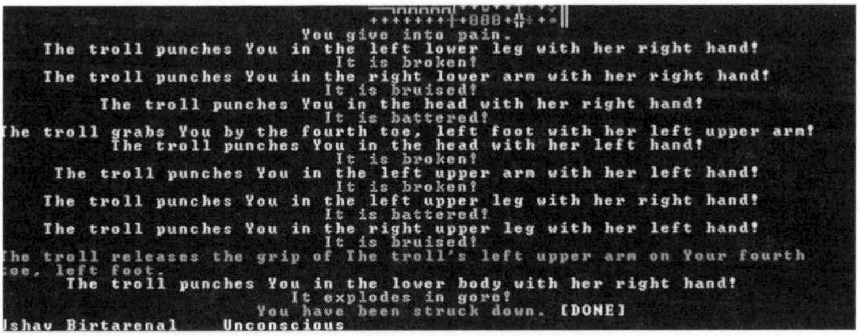

You give into pain.
The troll punches You in the left lower leg with her right hand!
It is broken!
The troll punches You in the right lower arm with her right hand!
It is bruised!
The troll punches You in the head with her right hand!
It is battered!
he troll grabs You by the fourth toe, left foot with her left upper arm!
The troll punches You in the head with her left hand!
It is broken!
The troll punches You in the left upper arm with her left hand!
It is broken!
The troll punches You in the left upper leg with her right hand!
It is battered!
The troll punches You in the right upper leg with her left hand!
It is bruised!
he troll releases the grip of The troll's left upper arm on Your fourth
oe, left foot.
The troll punches You in the lower body with her right hand!
It explodes in gore!
You have been struck down. [DONE]
shav Birtarenal Unconscious

Dwarf Fortress's realistic combat simulation,
down to modeling of individual toes

* As of 2016, he claims the game is 42 percent complete.

Dwarf Fortress takes the life-simulation approach the furthest of any game, and in doing so shows where such approaches break down. In particular, it illustrates the problems that can arise with computational models based on folk taxonomies. *Balance of Power*'s limitations are felt in the sheer number of factors left out (trade, treaties, religion), but the limitations of *Dwarf Fortress* emerge through the interaction of the innumerable factors it includes. The same complexity that gives convincing richness to the world also lets that world spin out of control past its creator's intentions.

Dwarves may be seized by the urge to create, leaving engravings on the wall for future explorers to find after a settlement has died out. The engravings, computed by algorithm from a fairly large set of possibilities, reflect the dwarf's psychological state and past experiences, such as a "masterfully designed image of a dwarf . . . The dwarf is screaming."

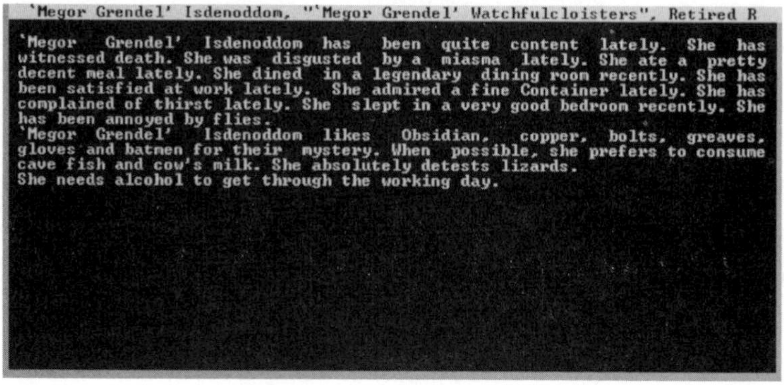

A procedurally generated engraving by a dwarf

The infamous chronicle of the "Boatmurdered" fortress tells of a miserable clan surrounded by wild and angry elephants that have a habit of killing dwarves. A group of players took turns running the fortress. Early on in the game, one player constructs a mechanical doomsday device that, when released, floods the entirety of Boatmurdered's surroundings with lava.* A subsequent player triggers this device in

* The player chronicled his year in the style of *Deadwood*'s Al Swearingen: "To begin with, all of our fucking workshops and trade goods are sitting outside in the fucking rain. One of the previous Overseers must have been some sort of shallow-dwelling skygazer because having our production out here is just inhumane to the poor hoopleheads who have to stand out there. I almost went fucking crazy having to be under the goddamn open sky for the whole trip out here, instead of in a nice safe cavern, and some of these poor bastards have been

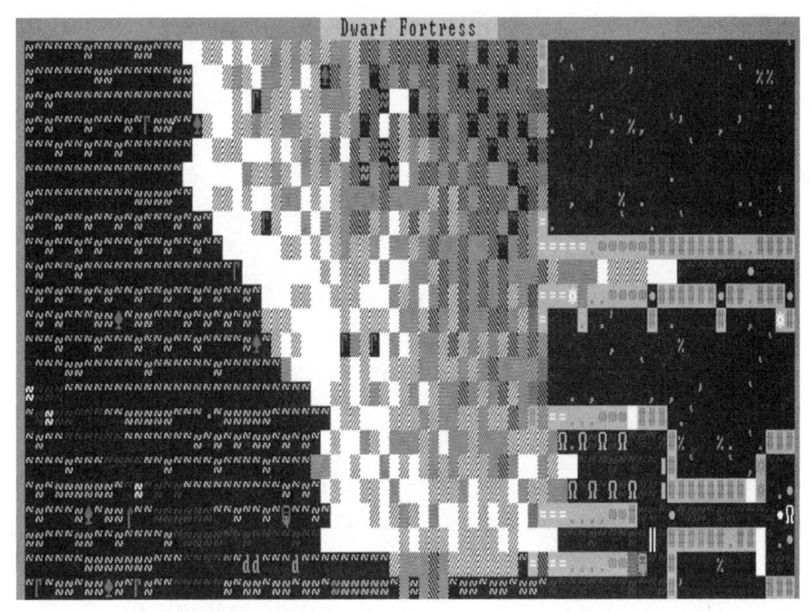

Lava meets the floodwaters, producing a white and gray cloud of steam and burning everything alive.

desperation after his dwarves accidentally break through the aquifer beneath the entire area and flood the outside terrain, in the hopes that the lava will evaporate the water. It does, but in doing so, scalds and kills elephants and dwarves alike, leaving a deadly cloud of steam and the purple miasma of decaying corpses.

All this activity plays out on top of *Dwarf Fortress*'s byzantine world-modeling algorithms. Adams's intent is for the systems of the world to be constructed *generally* rather than specifically. That is, rather than coding up the behavior of water and lava separately, Adams constructed a generalized fluid dynamics model for the game, from which the flow of any liquid under different situations of pressure could be derived. Likewise, weather is not just something that makes objects and creatures cold and wet, but emerges from an entire meteorological model that can produce fog, rain, snow, and all the major types of clouds, depending on the underlying conditions. These models collec-

working out in the open for four years now. Four years standing out in the rain, or even worse, under that horrifying yawning expanse of blue open sky."

tively create a believable, primitive environment in which each aspect is modeled with sufficient realism that the interactions between these systems are *also* realistic—most of the time.

Dwarf Fortress's sheer number of interacting subsystems can produce unpredictable and surprising results, as in this example:

> Once, a town's executioner had to execute a criminal with his teeth because he had lost both of his arms. He ended up biting the other dwarf's arm off and the command to spit the arm out never fired, so for years in-game the dwarf walked around with an arm in his mouth.

Execution-by-biting-limb is *not* a bug, since the criminal was indeed executed—only the failure to spit the arm out is. I can hazard one guess as to how this happened. The death sentence demanded that the executioner "attack" the criminal until the criminal was dead. The complex attack logic looked first to what weapons the executioner could use, but found that there were none because the executioner had no arms. The attack logic then turned to teeth and biting as an attack form, possibly seeing it as more deadly than kicking or head-butting. Yet after the executioner killed the criminal by biting the arm off, the executioner failed to spit the arm out because the situation was not typical combat. Normally, combat would result in other dwarves observing the fight, maybe getting involved, getting angry, etc. But in dwarf society, as in our own, an execution is a special case of combat in which others are not supposed to get involved. It is regulated by an implicit law.* Rather, executions were most likely designated as special case events, and certain mechanisms in dwarves that would *normally* cause them to react in violent ways were switched off. One of these disabled combat mechanisms, however, also affected whether the executioner would spit out a limb after biting it off—as outside of combat, a dwarf wouldn't end up with an opponent's limb in his or her mouth. So the special case produced a previously unknown bug. This is only my guess, however. Perhaps the arm simply wasn't coded as something that needed to be spit out.

* Law is not yet an explicit system in *Dwarf Fortress,* but it is on the to-do list. Adams: "Do we do law, property, status, customs, economics, boats? Which of those comes first? Because they influence each other so much, law influences economics, economics influences law."

Adams himself describes exactly such a case involving cats getting drunk and vomiting. This bug (what in software terms might be called a "corner case") demonstrates, acutely, how seemingly logical computer models can produce illogical behavior when joined together.

> I added taverns to fortress mode, so the dwarves will go to a proper establishment, get mugs, and make orders, and they'll drink in the mug. And, you know, things happen, mugs get spilled, there's some alcohol on the ground.
>
> Now, the cats would walk into the taverns, right, and because of the old blood footprint code from, like, eight years ago or something, they would get alcohol on their feet. It was originally so people could pad blood around, but now any liquid, right, so they get alcohol on their feet. So cats will lick and clean themselves, and on a lark, when I made them clean themselves I'm like, "Well, it's a cat. When you do lick cleaning, you actually ingest the thing that you're cleaning off, right? They make hairballs, so they must swallow something, right?" And so the cats, when they cleaned the alcohol off their feet, they all got drunk. Because they were drinking.
>
> But the numbers were off on that. I had never thought about, you know, activating inebriation syndromes back when I was adding the cleaning stuff. I was just like, "Well, they ingest it and they get a full dose," but a full dose is a whole mug of alcohol for a cat-sized creature, and it does all the blood alcohol size-based calculations, so the cats would get sick and vomit all over the tavern.
>
> People helped me with this. We were all looking and figuring out, "What the heck is going on here?", and that was the chain of events. It's like doing the detective work to figure out that entire chain of events is what happened. You can see how adding just a tavern that gave the opportunity for spilling alcohol, which was really uncommon before, now all the spilled alcohol starts to form in one location where something could start to happen. You activate bugs and little parts of code from eight, six years ago where you just didn't balance the numbers because it didn't matter. You don't want to spend time doing balancing that doesn't matter, because then you lose a couple of days doing something for no reason.

What's crucial here is the combination of meticulousness and intuition—the same combination evident in Crawford's book on *Balance of Power*. Adams is comprehensive to an extent that would make Gygax blush, and programming requires him to fix and flesh out his mechanics in far greater detail than Gygax ever had to, but many of those details are constructed off-the-cuff. This is not true in every case: in the instance of weather and fluid mechanics, Adams did a great deal of research into physical laws and how they should function. But in the cases of human and animal behavior in particular, Adams compromises. In the cat vomit case, Adams underspecified a number of commonsense mechanics. Each in the series of mechanics, from cats' paws soaking in alcohol to cats licking paws to cats throwing up drunk, makes sense in isolation, but when combined, they produce a ridiculous result. Adams describes the process as one of manually determining every possibility:

> There are so many interlocking systems now, and when you say, "interlocking system," I mean, sometimes it's these random events like the cat thing where it just happened to work that way, but more often the interlocking is you manually make the spokes and little gear teeth. That's what you're doing. You're adding each little one that fits in with every gear you could think of. There's more gears that would fit in too if you remembered they were there, and you just do as much as you can. Every system probably relates to every other system in some way. Those just haven't all been realized yet. . . . You don't have to find every connection but you just don't break the game so bad that it falls apart.

As a software engineer, I'm terrified by the complexity that Adams describes. Every programming instinct insists that systems be kept as simple and contained as possible, with interactions carefully delineated and constrained. I want to keep the whole system in my mind at once. I want there to be as few manually specified special cases as possible. I want things to be clean. But Adams's intent to simulate *the world* in as much complexity as possible runs straight up against that instinct. "Every gear you can think of" is never going to be all of them. There will always be holes, more than a single human can cope with. Little elisions add up over time, and eventually they *do* matter. The seams start to show.

Computer scientist Alan Perlis wrote, "Every program is a part of some other program and rarely fits." In other words, when folk taxonomies are elaborated and combined beyond their initial scope, they show their seams and begin to split open. This need not require contact with the real world; the simple interaction of independent taxonomies can produce new conflicts that would never have arisen independently. In the *DSM*'s case, the conflict between the competing demands on the *DSM* caused its directors to retreat from external requirements into confused and isolated rhetoric. Yet without the mathematical demands of computers to keep them honest, they only made their classification more private and inaccessible and arbitrary. Software engineers, however, control the computational meta-system by which the entire world is being quantified. Psychiatrists try to carve nature at the joints, but software engineers go beyond that. We carve the whole world at its joints, and we create tools that enable everybody to do the same. And then we program computers to do it for us. Computers increasingly assist us in role-playing ourselves.

Frances Allen conceived of the *DSM* as "a guidebook, not a bible—a collection of temporarily useful diagnostic constructs, not a catalog of 'real' diseases." And Dungeon Masters "use the rules only to help the players suspend their disbelief and to suggest new, better dramatic possibilities for your story." Such guidelines provide a precedent-guided restriction on complete arbitrariness, but allow for creative variation. Such modeling is a heuristic and humanist art, one that computers do not yet practice well, if at all. In *Dwarf Fortress,* by contrast, the rules are binding, not suggestive. As computers increasingly take over the role of referees from humans and tighten the enforcement of folk ontologies, what happens to our personae? Whether as doctor or Dungeon Master, computers do things differently. Computers make it much harder to revise the rules. And as systems reach and surpass the complexity of *Dwarf Fortress,* they increasingly become undebuggable, *evolving* unforeseen bugs on their own.

The first company to confront this dilemma was Google. I was there when they did.

INTERLUDE:

ADVENTURES WITH TEXT

Pico and Sepulveda
Where nobody's dreams come true

—EDDIE MAXWELL

HUMANS CAN FINESSE the gaps in their world models that computers cannot—with language. We inherit our language from our surrounding environments, and how we speak reflects the virtues and faults of the immediate culture around us. I grew up in the remote suburbs of Los Angeles in as close to a cultural void as could exist at the time. The limits of my world were local, hardly reaching beyond the space of a few dozen miles in any direction. Malls, bookstores, video rental places, and a small library. Separated from any sort of organic heritage or legacy, my family resided in a town that was crafted and manicured to evoke no one specific culture. Aside from the looming, monumental presence of the Pacific Ocean, I lived in a blank canvas on which no one was painting. I subsisted on UHF's television flotsam and whatever I found at the local library and B. Dalton Bookseller.

This was before the web. Culture as such reached me only through older media: books, movies, television, comics. Bookstores, libraries, and video rental stores were the only points of connection to worlds unlike my own. Computers were a part of my life, but there was little

The Gadget, the fission bomb that exploded during the Trinity test in 1945 (US DOD)

art in them. I remember *Balance of Power,* Logo, and the diabolical *Robot Odyssey,* the last being possibly the hardest computer game of its era. Such computer creations exercised my mind and spurred my curiosity, but they rarely imprinted themselves on my heart as many books and movies did. But there were a few programs that did. *Trinity* was one of them. Like *Balance of Power, Trinity* was a product of the nuclear fear of the Cold War, but where *Balance of Power* is strategic, *Trinity* is literary.

Author Brian Moriarty named *Trinity* after the site of the first atomic weapon test on July 16, 1945. The spirit of Lewis Carroll hangs over this 1986 text adventure, the caterpillar's mushroom transformed into the nuclear mushrooms that haunted the Cold War generations. *Trinity* was my introduction to the nuclear arms race. It opens with an atomic holocaust and the end of the world, then sends you back in time to the original Trinity test site in 1945 to stop it, via a fantasy looking-glass world connecting all nuclear detonations. I cannot quote you much from *Ultima* or *Might and Magic,* but I can tell you of the words I learned from Infocom games ("newel," "palimpsest," "affidavit"), as well

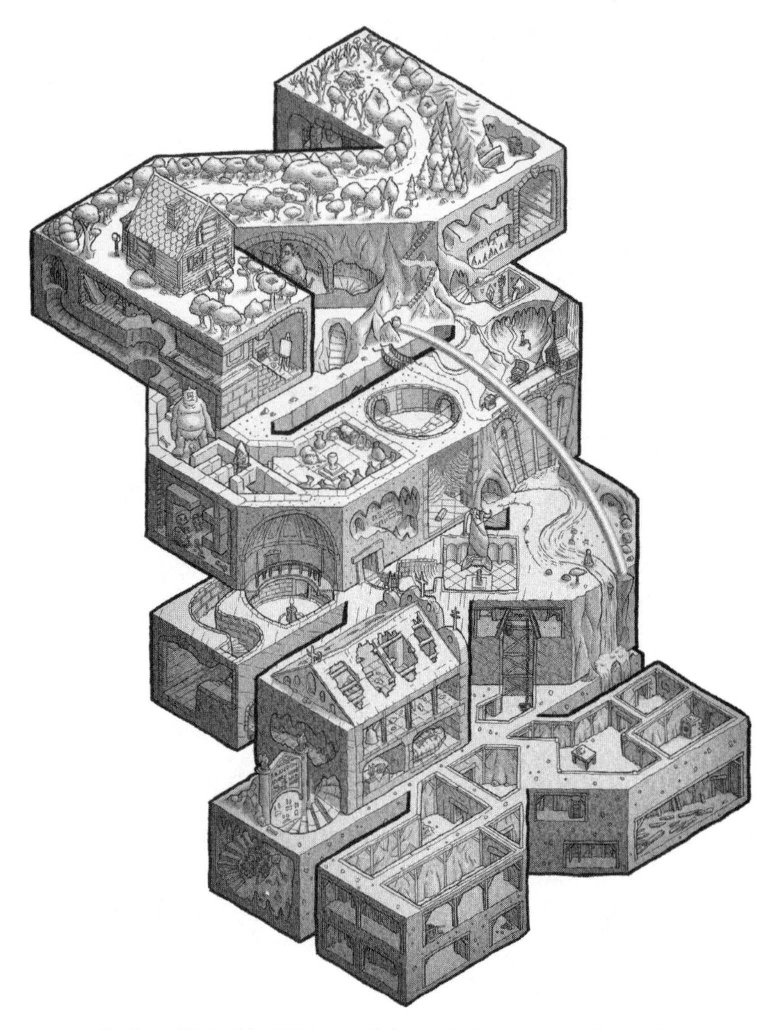

Zork, published in 1980, was Infocom's first game, authored by Marc Blank, Dave Lebling, Bruce Daniels, and Tim Anderson. This illustration by Laurence Jenkins contains every location in *Zork*.

as lines like the mantra of *Trinity:* "All prams lead to the Kensington Gardens."* *Trinity* still haunts me.

Players control text adventures through imperative commands. The game provides you with a description of a setting and situation, and

* The line is from J. M. Barrie's *The Little White Bird*. Beyond Barrie and Carroll, the game quoted Emily Dickinson, Herman Melville, and Alexander Pope.

you issue an instruction for what you want your player character to do. By convention, the game speaks to the player in the second person and present tense. In the imaginary transcript below, the player's commands are the lines beginning with an angle bracket in bold type, the computer's output in regular type.

THROUGH THE KLEIN BOTTLE
An interactive nostalgia trip
Release 1 / Serial number 170221

You find yourself in a strangely familiar unfamiliar place.

Sculpture Garden
You are standing before the grandest sculpture in an intimate yet sweeping garden. The sculpture arches over your head, tangling itself into twisted shapes somewhere between Calder and Tinguely. The metalwork curves so much that it is difficult to tell where the inside ends and the outside begins.

Corrugated steps rise up into the sculpture to the north and south.

A baby grand piano stands between the two entrances. Its lid is unusual. A folded paper card sits on the keys. The words *Felix Klein 1849–1925* are engraved on the piano.

> who am I?
You are David Auerbach, or at least you were the last time you checked.

> how old am I?
Your age is strangely indeterminate. You don't feel like a child, as you appear to possess the memory of decades behind you, but simultaneously you look at yourself and see the gangly limbs and stiff clothes of your twelve-year-old self. The strangeness of the environs makes it that much harder to locate yourself.

> read the card

You take the card from the keys and unfold it. *"Please play Bach's musette,"* it reads. As if on cue, the simple melody of the piece rings out on chimes from somewhere inside the sculpture.

> play musette on piano

You haven't played it in over twenty years, but the finger movements come back to you immediately. You begin to play, but recoil from the dissonance. The piece is all wrong. Or rather, the piano is all wrong. The keys are reversed, so the lowest notes are at the right end of the keyboard and the highest at the left.

> play musette reversed on piano

You try to flip your finger movements to match the inverted layout of the keys, but it's no good. Your muscle memory does not admit to such geometrical transformations.

> kill me

You're indispensable.*

> examine the piano

It's backward, for one. Also, the lid is not flat, but instead takes the form of an impossibly curving piece of wood in a vague jug shape. The prop stick appears to warp straight through the lid, making it difficult to tell where the outside of the "jug" ends and the inside begins.

The words *Felix Klein 1849–1925* are engraved on the piano.

> go north

You ascend the north side of the sculpture.

* This wonderful line is borrowed verbatim from *Trinity*. Typically, other games obeyed the command, responding, "If you insist" (as in *Starcross*) or simply "Done" and ending the game.

North Sculpture
The "floor" of the sculpture curves up and around so as to make you dizzy. Your footsteps do not seem to be taking you in the direction you think they should be.

You feel certain you have been here before.

> go up
The floor twists alarmingly as you continue your ascent.

Top of Sculpture
The metalwork clears a bit here, yet the arc of the structure is as tangled and incoherent as ever. You feel all turned around.

The tunnel of the sculpture curves down to the north and south.

Your sense of direction is very confused, but your mind has snapped into focus. You are in the Klein bottle from *Trinity*. It is the four-dimensional analogue to the Möbius strip. Where the Möbius strip turns its two sides into one by twisting into the third dimension, the Klein bottle turns two surfaces, inside and outside, into one single surface by twisting into the fourth dimension. In *Trinity*, it was a large topiary that formed the Klein bottle rather than a metal sculpture, but the shape is the same.

> remember trinity
You think back to Brian Moriarty's 1986 text adventure *Trinity*, a haunting game about nuclear Armageddon. It's imprinted on you indelibly. You remember the frightening moment of meeting the old Japanese woman in Kensington Gardens at the beginning of the game, just before World War III starts and the world ends.

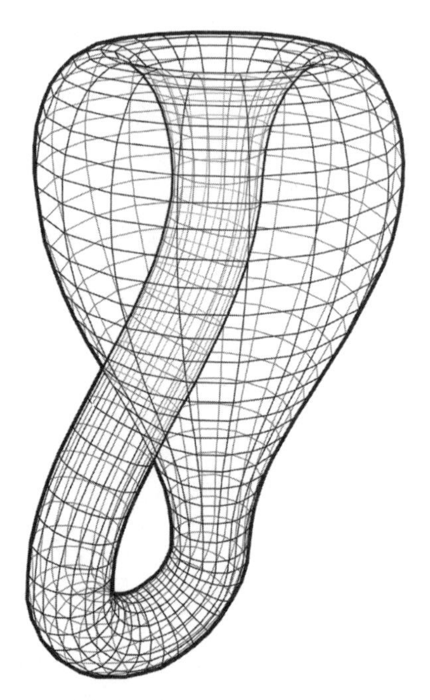

The one-sided Klein bottle. Since it does not actually intersect itself, it cannot be realized in three dimensions. Four dimensions are required.

> You begin to approach the old woman, but stop in your tracks. Her face is wrong. You look a little closer and shudder to yourself. The entire left side of her head is scarred with deep red lesions, twisting her oriental features into a hideous mask. She must have been in an accident or something.

Your character will later travel back in time to Nagasaki and meet her as a child just before the bomb hits. It was a devastating game. It provided a new emotional venue, a new context for you to experience a half-mimetic, half-fantastic world as a shadowy player-character whom you inhabited.

> go south
South Sculpture
The "floor" of the sculpture curves up and around so as to make you dizzy. Your footsteps do not seem to be taking you in the direction you think they should be.

> go down
The floor twists alarmingly as you descend.

Sculpture Garden
You are standing before the grandest sculpture in the garden. It arches over your head, tangling itself into twisted shapes somewhere between Calder and Tinguely. The metalwork curves so much that it is difficult to tell where the inside ends and the outside begins.

Corrugated steps curve up into the sculpture to the north and south.

A baby grand piano stands between the two entrances. Its lid is unusual. The words *5291-9481 nielK xileF* are engraved on the piano.

> examine the piano
The lid is as confusing as before, but the piano is now forward. By walking through the surface of the Klein bottle you have effected a mirror-reversal on yourself. But as far as you're concerned, you have mirror-reflected the world.

> play musette on piano
You warily touch the keys, but find them to be properly oriented now. The charming melody of the musette, still so familiar to you, plays out on the piano easily.

The simple melody resonates against the metal sculpture. Each note seems to echo a resonance frequency deep within the sculpture, until the entire structure is vibrating uneasily above and around you. As you finish the piece, some of the girders are visibly undulating, and the screams of metal against metal set your teeth on edge. Then with a huge clamor, the structure collapses all around you, miraculously leaving you unscathed, but hitting the ground with the titanic force of a building being demolished.

When the dust clears you are somewhere else.

Ruins
You are outside, alone, in a pile of rubble and destruction that extends as far as you can see. But that is not terribly far, as smoke and clouds of ash are everywhere, and your line of sight is interrupted every few feet by a fire, a pile of detritus, or simply a huge, black, burnt-out slag.

> call my wife
(taking your phone out of your pocket)
You have no reception.

> look at phone
You have zero bars. Either you are out of range, or the destruction is blocking the signal.

> look at date on phone
It is September 11, 2001.

Seeing the date gives you a stronger sense of identity. You are only a few years out of school, working at Microsoft. Free of the nuclear paranoia that fueled *Trinity*, you lived through the odd period of stasis in the 1990s after the collapse of communism, when neoliberalism was ascendant and the problems of the world were very far from the problems of the United States. Your ignorance was shattered on this particular day, and you would spend years coming to terms with the world as it really was rather than how it had seemed to be from your partial and privileged vantage.

And you recognize this to be a dream that haunted you occasionally during those years, in which you found yourself right inside the site of the demolition of the myth of being posthistorical and enlightened.

> wake up
Done.

```
**** You have returned ****

In 18 turns, you have achieved the rank of Mnemonist. Would
you like to start over, restore a saved game position, undo
your last turn, or end this session of the game?
(Type RESTART, RESTORE, UNDO, or QUIT)

> quit
```

Moriarty's passage about the Nagasaki survivor chilled me like little else in computer games. In the depiction of a single, nameless character, whom the player sees first as a scarred, elderly woman, and later as an innocent child prior to the Nagasaki bombing, the collective weight of fifty years of nuclear anxiety and trauma fell on me. It hit my brother even worse. Only eight years old at the time, he took a look at the game and was seized with the fear of nuclear war. I understood his fear. There were no bad guys in *Trinity*. It was a portrait of a world on the precipice, having discovered the secret of its own annihilation. And my character in the game walked through it and saw the nightmare. The old woman was the crux: the terrifying power wasn't a hypothetical, future doom. It was already here, and it already had taken victims. That passage awakened me, much as September 11, 2001, did over ten years later.

Trinity's text was written by a human, not a computer. How did its placement in a text adventure affect its impact? Text adventures graft human language on top of algorithms. Instead of regimenting human life into numbers and equations, they refuse automation. In this regard, text adventures are the opposite of role-playing games, their complement and counterpart in the history of computer games. While RPGs put your character's statistics at the surface, text adventures only sparsely quantify their worlds. There are no explicit statistics or dice rolls in adventure games. Ironically, while the cognitive biases of RPG designers get baked into RPG algorithms, text adventures isolate these biases at the level of *human language,* leaving the underlying plumbing mostly free of the quantifying models of life. *Exodus* and *Wizardry,* both of which effectively port the mechanics of Dungeons & Dragons wholesale to computers, are comprised of numbers and maps and diagrams. So were computer wargames. Text adventures foreground the dialogue between creator and player. The human and computer layers are distinct.

These points apply to adventure games more generally, including those with graphics, but I wish to focus on the tradition of text adventures, as pioneered by *Adventure* and *Zork*, for two reasons. First, they attempted to deal with the algorithmic problem of understanding human language in a primitive yet ingenious way: they let players *type* what they wanted to do *in English*, rather than having them select words from a menu or push a button. Second, they showed me how computers could provide immersive and moving narrative experiences, integrating the most ineffable parts of human creativity to do so.

At the time, it was marvelous to behold that a computer could *understand* the text I typed into the computer, including complicated constructions with adjectives and prepositions. In this Infocom was peerless: their parser was far more sophisticated than that of most games, which had trouble getting too far beyond the basic VERB OBJECT (GET GOLD, KILL TROLL, etc.) command form. But Infocom built a general-purpose parser that they employed and refined. Its ability to handle and interpret a variety of expressions remains an outstanding technical achievement.

The infamous Babelfish puzzle in Infocom's game *The Hitchhiker's Guide to the Galaxy* asks the player to obtain a universal translator fish from a dispensing machine, but all sorts of devices and furniture get in the way of obtaining one. The puzzle is ridiculously hard. When I went as a kid to get my copy of *Dirk Gently's Holistic Detective Agency* signed by Douglas Adams, I overheard Adams telling the person in front of me *how to solve that puzzle.** It was that difficult, and there was no internet available for hints. (Sales of hint books were a major revenue source in the late 1980s.)

Yet despite the fiendishness of the puzzle, the parser doesn't impede the player. The problem is figuring out what to do, not figuring out how to express it to the parser. That may not come as much comfort to players bashing their head against the wall trying to derive the next

* Douglas Adams exerted a greater influence on me than I realized at the time. The mix of science-fiction clichés with acidic British irony fit me quite well in my middle-school years, and served as a stepping-stone to Kurt Vonnegut, who in turn was a stepping-stone to more "serious" and "literary" fiction.

step in the Rube Goldberg machine to obtain the Babelfish, but it is a testament to the strength of Infocom's parser (and its playtesting). Interactive fiction guru Graham Nelson uses this particular puzzle to explain the game's immersive interaction: "Without puzzles, or problems, or mechanisms to allow the player to receive the text a little at a time—whichever phrase you prefer—there is no interaction." He gives a sample transcript of the beginning of the puzzle:

```
>examine machine

The dispenser is tall, has a button at around eye-
level, and says "Babel Fish" in large letters. Anything
dispensed would probably come out the slot at around
knee-level. It bears a small label which reads "Another
fine product of the Sirius Cybernetics Corporation."

>press dispenser button

A single babel fish shoots out of the slot. It sails
across the room and through a small hole in the wall,
just under a metal hook.

>remove dressing gown

Okay, you're no longer wearing your gown.

>hang gown on hook

The gown is now hanging from the hook, covering a tiny
hole.

>push dispenser button

A single babel fish shoots out of the slot. It sails
across the room and hits the dressing gown. The fish
slides down the sleeve of the gown and falls to the
floor, vanishing through the grating of a hitherto un-
noticed drain.
```

```
>put towel over drain

The towel completely covers the drain.
```

This is a best-case scenario. The player is more likely to spend time examining random objects, trying to place different things on the hook and over the drain or engaging in crazier antics like pulling the hook out of the wall or pouring things into the drain, before stumbling on the solution. Whether the puzzle is good is a subjective matter, but the parser does a passable job of indulging the player's efforts and offering *meaningful and relevant* responses to the player's attempts. The game walls off possibilities that might have worked in real life. Yet the text parsing is *sufficiently* good that some segment of players felt the puzzle was "fair," the ultimate test of the quality of a game.

By 2000, an Infocom-derived parser (in the TADS language) became sufficiently sophisticated and robust to allow for a game like Andrew D. Pontious's *Rematch,* a single-move game that takes place in a pool hall, just before a car crashes through the window and (usually) kills you. Its Rube Goldberg mechanism fires in response to this single command, which wins the game:

```
>whisper to Nick to dare Ines to throw page 351 with cueball
at loudmouth
```

The ensuing chaos caused by this command results in you, Nick, and Ines getting out of the way just before the car crashes through the window. As with the Babelfish puzzle, the mechanism is diabolical to discover, requiring repeated attempts and deaths, but *expressing one's intentions* never stands in the way. The parser is that strong, though when I first played it I could count the seconds as the computer ground away trying to process the "whisper" command.

An unfair puzzle requires verbiage so rigidly and arbitrarily constrained that the player is frustrated trying to reach it. In these cases, a player has to guess at the hidden logic in the code itself rather than the overt logic presented in the text of the game. In other words, the player has to read the author's mind. A classic example is from the charming yet flawed 1986 game *The Pawn,* by Rob Steggles and Anita Sinclair, in which the player must plant cannabis: the correct command is "PLANT THE POT PLANT IN THE PLANT POT WITH THE TROWEL." But slight variations, from "PLANT PLANT" to "PLANT PLANT IN POT" and beyond, yield completely unhelpful responses like "I don't follow you." Only a command that is very close to the initial, awkward phrasing works. The game trips over itself when it has to distinguish between multiple senses of "pot" and "plant." The computer is trying to speak a foreign language, but reveals its lack of familiarity with how English functions. Unless you know the phrasing that the author wants you to use, your chances of guessing are slim indeed. In letting the limits of its code show, the puzzle fails.

What's going on here bears *no resemblance whatsoever* to how humans use language. Infocom's parser does not "understand" the player's commands in the way that humans understand each other. But it does a good job of *seeming* to understand, and that feat of prestidigitation is at the heart of how computers have become more human. We buy into the tricks that programmers play to make computers seem more human, until those tricks slowly *become* real.

At their best, these tricks in text adventures are in the service of art. I did not feel tricked by *Trinity*'s parser any more than I felt tricked by Kōbō Abe's or Virginia Woolf's symbolism. The parser and its primitive yet skillful understanding of human language help the player identify with the protagonist of the game in a way that is different from the experience of reading fiction. Some authors of text adventure games were attracted to the possibilities of player interaction: in addition to Douglas Adams, Robert Pinsky contributed his whimsically surreal prose to *Mindwheel* and Thomas M. Disch wrote the self-reflexive noir of *Amnesia*. The degree to which text adventures' novel yet limited interactivity was utilized varied. Sometimes it was little more than mere branching narratives of the sort employed by John Fowles's *The French Lieutenant's Woman* or *Choose Your Own Adventure* books. But despite text adventures' limited understanding of language, the ability to type in a command rather than select from a list gave a greater impression of involvement and possibility. A world could be parceled out in small chunks driven by the wandering eye of the reader. When, as the player character of *Trinity*, I encountered the old woman and read "*Her face is wrong*," I felt a visceral immediacy. I was not reading about someone encountering a survivor of Nagasaki, but felt that I myself was staring at her with a mixture of shame and curiosity. It was in part from the new modality of interaction that these games were able to achieve their emotional impact. Yet these effects were achieved in the absence of *computer understanding*. The meaningful content still existed separately from the mechanical code executing underneath it. It was through the *medium* of computers, not through computer *intelligence,* that authors like Brian Moriarty succeeded at their chosen art.

At their worst, the poorer Infocom games and the vast majority of non-Infocom text adventures were too circumscribed in their possibilities, too inept in their understanding, or too illogical in their construction. It was sometimes impossible to form a smooth bond between player and game.* When *Madventure* told me that I had come to a fork in the road, I needed to *take the fork* in order to progress. This was less interactive than irritating. In the 1980s, though, even very flawed games still held a certain magic over players, because the nature of that interaction was so new.

The question for companies like Google and Facebook became, how intelligent could computers *seem* to be in the absence of truly *being* intelligent?

* Douglas Adams's second collaboration with Infocom and Michael Bywater, *Bureaucracy,* made a virtue out of this fault by fighting the player at every turn, replacing the traditional player score with a measure of the protagonist's blood pressure, which started at 120/80 and shot up every time the player was vexed by the game—say by using a word the game didn't understand. Like most text adventure diehards, I was more amused than annoyed when my character *died* of a heart attack because I had made a typo.

PART III

7

BIG DATA

From the Client to the Cloud

> Men build their cultures by huddling together, nervously loquacious, at the edge of an abyss.
>
> —KENNETH BURKE

WHEN I WAS a child, code thrilled me with its elegant power to perform near miracles with a few lines of Logo. I was amazed that one could do so much with so little. Google, in its early years, performed the same sort of magic, but with data. Google amassed an unprecedented amount of data about the web, then developed the simplest and most elegant methods for analyzing it.

Everything was bigger at Google than it had been at Microsoft. When I arrived at their Mountain View campus in 2004 as a journeyman software engineer, I was surprised at how differently things worked there, and how differently I worked there. I had coded servers at Microsoft, but Microsoft's entire business was based around the PC desktop. That focus was one of the primary reasons why Microsoft found it hard to shift into the internet age, where the locus was not a home or workplace PC but a web server or database somewhere else on the internet—what's today known as the "cloud," the amorphous mass of data that floats around and above us, only dimly visible. Unlike Microsoft,

Google knew servers from the beginning. We worked on Linux, the free operating system based on Unix that had been uniquely designed for networked computers in the first place, rather than retrofitted to it as Windows had been. All engineers shared a single enormous code repository, rather than having a different repository for each team, as Microsoft had had. We had machines at our disposal: thousands upon thousands of them, in massive clusters around the country and eventually around the world. These machines were always running, and they were always running my code. Automated test infrastructures made sure that my code changes didn't break existing functionality—and automatically sent nagging emails to me, my manager, and lots of other people if I did. It was amazing.

At Microsoft, I tested the Messenger server by running a build of it on a few local machines. There were a few dozen test servers available for shared use. It was quicker to launch it on my own work machine. Microsoft's first real excursion into web services had come with the purchase of Hotmail, which had their own home-brewed mechanisms unsuited to general-purpose usage. Microsoft took years to build out a robust server infrastructure. Here, Google was a decade ahead of Microsoft.

It helped that I was fortunate enough to work with some of the best software engineers I'd ever met, and in particular under a technical lead, Arup, who was one of the most careful and comprehensive people I'd met in any field. He had a preternatural capacity for anticipating anything that could possibly go wrong and making sure our team handled it before it did go wrong. It made for fewer emergencies than I'd ever encountered in engineering, and remarkably low overhead, as our team of ten tended just to communicate informally rather than sitting down for long meetings. I learned that a small group of high-quality coders with a top-notch lead could accomplish vastly more in a month than some of my old teams had in a year.

The more significant difference was data, and how much of it there was at Google. While Microsoft had succeeded as a *software* company, Google's lifeblood was *data*. Google needed software to collect, store, and manage this data, but at Google, *software served data*. The advent and exponential growth of the web, which was reaching hundreds of billions of pages by the mid-2000s, required that there be an organized, comprehensive system to fetch, analyze, and retrieve that data en masse and at top speed. Google was the first company to do this. In the 2000s, Google came to own data in much the way that IBM owned the mainframe, Microsoft owned the PC, and Apple owned the mobile device.*

At Google, I was able to command a thousand machines at the push of a button and analyze billions of web pages in minutes. At Microsoft, the data carried by the Messenger servers, even at peak volume, was fairly small. A few million users sent a few million messages every minute. This was manageable on a few dozen servers. But a corpus of hundreds of billions of web pages was far beyond not just what any one human could sift through, but even what any one computer could sift through. Analyses were performed through massive parallelization and partitioning of data, in order to produce statistical breakdowns and data reductions that could then be used to return relevant search results. If I wanted, say, to know the most popular words used on web pages, then I'd send a couple billion web pages each to a thousand servers. Each server would analyze its portion of data, then pool the results to be analyzed on yet another machine. This chained process of quantitative analysis was central to Google's operations, and it became central to my life for the years I was there. It was also beautiful.

While Google even then made use of machine learning mechanisms, most of Google's operations, and certainly the areas I worked in, avoided "deep learning" in favor of raw number-crunching and dumb analysis. By "dumb" I mean that rather than trying to infer anything about the "meaning" or "structure" of the data, the general approach at Google was to see how much could be gotten out of what little explicit structure was there. In the early days circa 1998 to 2005, Google used word frequency and PageRank to see which pages linked to where. Soon, however, search engine optimizers (SEOs) tried to game Google search in order to promote their pages as highly as they could. Google, in turn, modified their algorithms to try to reinstate fairness, in what was termed the "Google dance"—because the order of results for a query would reshuffle. But initially, Google's approach was austere: maximize the quality of the search engine by sticking to the raw data—the link

* Some may suggest that Facebook belongs in this list in "owning" social, but I persist in believing that they are another data company on the model of Google; they have just used very different methods to acquire that data and have shifted the locus of data to *people*.

graph, word occurrence, co-occurrence, and order, and some amount of HTML page structure.* Google could *not* be certain of a page's subject matter, the meaning of a piece of text, or the quality of that text; these are matters that even humans could disagree on. These were relevant questions, but they were to be addressed only *after* Google had squeezed as much as it could out of the "dumb" data, because they were much harder problems. As ever in software engineering, Google picked the low-hanging fruit first. As Google research director and AI maven Peter Norvig said, "Simple models and a lot of data trump more elaborate models based on less data."

The goal of a search engine, and of information retrieval more generally, is to take a large amount of content and allow users to issue search queries (like "best camera" or "who is the president?") that return the most relevant results from within that content. There are three main components to any search engine: crawling, indexing, and ranking. The crawler, or webspider, is responsible for obtaining the content in the first place, by systematically downloading every web page it can find—billions of them. The indexer then takes these pages and breaks down their content, creating keyword lists through which pages can be located for search terms, as well as any other information that might give an indication of a page's value and relevance. It creates what is effectively a gigantic database of the web. Finally, the ranker uses the indexer's database in order to determine the relevancy of a page relative to a particular search query, and is responsible for the order of the search results that a user sees.

Google's search engine was comprised of a pipeline of servers that worked rather like an assembly line to crawl and index pages. I worked on the crawl. Our mission was to collect all the pages of the web, extract the links from those pages, find all the *new* pages in those links that we hadn't yet crawled, and then start over again as soon as possible with an updated set of pages. Since web pages frequently change, we recrawled sites as often as deemed necessary. This process became increasingly frequent, and some webmasters wrote in to complain that Google was using up too much of their bandwidth. This was especially a problem

* Google did build clusters of related words, such as "Jaguar" and "car" on the one hand and "jaguar" (noncapitalized) and "tiger" on the other. Yet these associations were made with as minimal a grasp of context as possible—simply by pure proximity.

for websites that had a lot of useless pages. Deciding which pages were useless—that is, which were of no interest to Google—became one of my main challenges.

Deciding if something was "useful" or "useless" was about as far as Google got when it came to understanding the significance of the content it indexed. A computer does not grasp what "useful" means as a concept, but Google had several wonderfully unambiguous metrics to gauge the "usefulness" of a piece of content, based around two core concepts: popularity and uniqueness.

1. **Who linked to a piece of content?** The initial version of Google in 1997 analyzed which web pages linked to other pages, and how they did so. If a million pages linked to a page on orange juice with the anchor text "orange juice," it was a reasonably good bet that that orange juice page was fairly useful to people who wanted to know something about orange juice. Whether that page was authoritative was another matter entirely, but Google could securely know that it was *popular*.
2. **Who clicked on the content?** Google tracked which results people click. Popular search results get boosts so that they become even more popular. High-ranking search results that are never clicked on get demoted. This is a feedback mechanism: the system "fixes" itself in response to the reactions of its users. Users provide this feedback without even intending to; by choosing the link that looks the most useful, they help Google learn how good its search results are. This assumed, of course, that users themselves knew what the best links were. Google functioned simultaneously as a *mirror* of people's tastes and as an *arbiter* of them.
3. **How similar was the content to the content on other web pages?** The rise of electronic media made copying trivial, because all it took to duplicate a piece of content was to make a virtual duplicate, which could then be shared

however you wished. With the internet, sharing with others around the globe became as easy as copying. The music and film industries discovered this when Napster and BitTorrent arrived on the scene. More innocuously, however, the internet caused there to be multiple copies of tons of reference works and boilerplate. Entire database manuals, technical documentation, public domain works, and more were duplicated across multiple sites. The more a piece of content was replicated across the web, the less likely it was that *that* particular copy was useful. At best, one copy was useful, which could be identified by seeing which copy was most popular among links. Often, however, such duplicated content was of little intrinsic value. It was cruft, boilerplate, and other flotsam and jetsam left over from the workings of various software packages. High-quality content tended to be reproduced *less,* not more.

4. **What was most distinctive about the content?** The least common words on a page generally were those that were most important. The word "person" would appear on billions of pages, but "mesothelioma" and "asbestos" appeared on far fewer, and on those pages "mesothelioma" and "asbestos" were far more relevant to the content than "person." By identifying the rarer words, Google could pin down more of the context in which a page was useful.

It was through these questions that Google not only discovered how to make the best search engine, but also how to make money from its search engine: through advertising. In 2000, Google found their golden goose when they released their AdWords service, in an internet monetizing coup that has not been since replicated, not even by Facebook. Unlike banner ads, which had low click-through rates, Google presented ads to users that were far more likely to be clicked. This owed to the unique position Google users were in: they were searching for a particular term—often because they were interested in *buying* something related to that term. If third-party advertisers could *bid* on showing ads for specific keywords, Google could show those ads at exactly the right time: namely, when users were searching for those keywords.

And advertisers were willing to bid up some keywords quite high. One of the most reliably expensive keywords over the last fifteen years has been "mesothelioma," bid on primarily by law firms looking for asbestos victims. Many law and insurance keywords go for over a hundred dollars per single click. In 2016, "best mesothelioma lawyer" cost you over $900 per click.

SEOs engaged in a never-ending effort to rig Google's results by trying to bias these factors into favoring their pages. Snagging a top Google placement in search results was valuable, and each of the four measures of usefulness above could be manipulated:

1. **Boosting links:** SEOs set up their own sets of "link farms" to artificially increase the incoming links to a page they wanted to elevate in Google's rankings.
2. **Boosting clicks:** SEOs set up bots to search for terms on which they wanted their pages to be found, then repeatedly clicked on their pages in the results.
3. **Artificial differences:** SEOs generated multiple copies of pages that were different enough not to be identifiable as related.
4. **Keyword manipulation:** SEOs stuffed pages with *other* rare keywords to maximize attention given that page so that people would see it in more search results than if the page were genuine content.

As SEO grew, Google devoted as many resources to fighting back against the attempted manipulation of its results as it did trying to improve the quality of the results—since the two tasks were effectively the same. SEO efforts made up a tiny fraction of the total content of the web, but because of Google's privileged place in the 2000s as the overwhelmingly dominant search engine, SEOs fixated on identifying Google's weaknesses and abusing them, in order to elevate their content over unoptimized content. By 2010, the original ranking recipe that had allowed Google to be so successful had become untrustworthy. Google frequently altered its recipe to utilize a different mix of criteria.

People like to speak of Google's ranking algorithm as its "secret sauce," but this ever-shifting recipe is one that not even Google's own engineers could write down. "Usefulness" had metastasized from a humanly comprehensible algorithm into an arcane calculation involving hundreds of individual factors per page. The relationship of Google's measure of "usefulness" to *our* measure of "usefulness" remained more or less steady, but this continuity masked how much Google's algorithm diverged from intuitive human thought processes. In the cloud, unlike in the world of the PC, there was no way to undo the complexity, as Microsoft had been able to with Windows 2000 (which finally fixed up the messy legacy left by MS-DOS and Windows 95).

Hangman

> When it breaks, you build it again. . . . Gotta fix it faster.
>
> —Brian Edison in George Armitage's *Hot Rod*

Here is an example of how useless content caused trouble for Google. Sometime around the turn of the century, someone wrote a notorious web implementation of the child's game Hangman, where each guess of a letter took a user to a new page with a unique URL that included all the letter guesses that had been made, like this:

Player Guesses A: http://www.mysite.com/fun/hangman/a
Player Guesses E: http://www.mysite.com/fun/hangman/a/e
Player Guesses I: http://www.mysite.com/fun/hangman/a/e/i
Player Guesses O: http://www.mysite.com/fun/hangman/a/e/i/o

By the end of a game, if the word was COMPUTER, the player might reach this "page":

http://www.mysite.com/fun/hangman/a/e/i/o/u/c/m/p/t/r

The entire alphabet can follow at the end of the URL, in any order. The game ends when the player runs out of guesses or completes the word. If there was a ten guess maximum, there would be 26!/16! distinct URLs containing unique permutations of ten letters—that's almost twelve trillion. By default, Google would try to crawl them all and fall into the quicksand of Hangman.

Hangman is an example of just how imperfect and ultimately *human* the web is. It may sound like a mechanical exercise to grab all the pages on the web over and over, but the web was varied and irregular from the very beginning. Web pages were not manufactured in some central factory by cookie cutters. Even today there linger all sorts of shims and spandrels that a handful of web developers thought were a good (or expedient) idea at one time or another. Some Google could just ignore; others, like Hangman, we had to deal with. Data is very rarely as neat as we imagine.

Google initially handled Hangman in a very inelegant way. We hard-coded a check for "hangman" and a few variants thereof that would stop the crawler from crawling those trillions of generated pages. This sort of special-casing is frowned upon, since all it took was calling the page "hagman" instead of "hangman" in order to defeat the check. One of my later tasks was to automate detection of this *sort* of crawler black hole, but Hangman in particular was so unusually awful that Google's engineers had made a special-case exception for it years before.

Even when you *think* you need to know nothing about the data you're dealing with, the implications of its content can sneak in. To Google's code, Hangman was more "interesting" than the average website or blog, simply by virtue of it being a special case. Yet these pathological cases were utterly marginal in a practical sense. Hangman signaled another division besides *useful* or *useless*. In a world where there is an overwhelming surfeit of data, more than we can process piece by piece, all data is either *representative* or *pathological*. Ordinary or bizarre. Standardized or broken. The data was either part of the machine, or a virulent entity that the machine couldn't handle.

I enjoyed these sorts of problems. I was looking for generalized solutions for a multitude of particular pathological causes. The web was not perfect, because humans were not perfect, but Google's code could be improved and tuned in order to cope with the imperfections of the web in better and better ways. This process can be frustrating; imagine engineers who work on ranking and their task of keeping results as relevant and high quality as possible. With the web constantly evolving and growing, maintaining the current standard of search performance becomes an exercise in running to stand still. Stop working on Google for a month or two, and the search engine would be far worse when you came back to it. The implicit challenge became one of trying to *outrun* the changes to the web—the search for algorithms that were sufficiently general and smart that they could handle whatever surprises the web might next throw at them. Google might need a new handler in order to understand content presented on web pages in the new HTML5 specification, but with luck and planning, the underlying indexing format was robust enough that it could still process the new data obtained from HTML5 without too much adjustment.* Or perhaps ranking algorithms are sufficiently robust that some new SEO tricks aren't able to foil them.

The dream, then, is to create algorithms of maximum generality while sacrificing as little specificity as possible. In reality, software engineers always compromise, blocking off a certain problem area and finding an expedient path that's neither too ambitious nor too kludgy.† And that sort of heuristic optimization is not so far off from what we do every day in life and language. We lump things into human-comprehensible categories while trying to respect special cases, but we never reach perfect abstraction, nor perfect comprehensiveness. It was working at Google that helped me understand this.

The Library of Babylon

> I have made a heap of all that I could find.
>
> —NENNIUS, *Historia Brittonum,*
> as quoted by David Jones, *The Anathemata*

I wrote a blog while I was working at Google—a "litblog." I wrote about deeply unpopular books and chronicled my reading of Proust. On my

* I wouldn't know. I wasn't at Google by the time of HTML5.

† A kludge is something like the Hangman special case: ugly, nongeneral, but effective for the time being.

website stats, I saw Google inhale my blog pages every few days, and spit them out in search results. But from work, I could see my site from Google's perspective. More accurately, I *didn't* see it. There was nothing special about my site, nothing that caused it to stand out. My pages were a couple hundred out of billions. They were typical, standardly formatted pages about esoteric subjects, to be indexed and retrieved in response to keywords like "Proust" and "Krasznahorkai" and "modernism." I was not tempted to rig Google to favor my pages. Even if I could have (which I couldn't), there was very little *to* rig. My pages showed up when people searched for my subjects, and didn't otherwise. The hot topics were locksmiths (lots of scammers) and mesothelioma (lots of ambulance chasers). Even if my blog had been artificially inflated in its rankings, that wouldn't have generated a great deal more interest in it. Google Search could skew people's attention, but not create it.

The god's-eye view offered by Google was ultimately indifferent to what went on outside its server farms. Every individual thing was too small to matter. Large sites like CNN or the New York Times or Wikipedia needed special care and attention due to their size and popularity, but those sites were not any more or less interesting to Google, merely more time-consuming. Content that was not pathological like Hangman was mostly meaningful as a representative of some more general type of content: news, image, video, or otherwise undistinguishable text. To Google, I was one indistinguishable segment of a very, very long tail. Google assimilated everything, and none of it meant anything.

In the Total Perspective Vortex, from Douglas Adams's *The Hitchhiker's Guide to the Galaxy* radio series and novels, an unlucky victim sees the vast expanse of the universe as well as a tiny marker that says "You are here." The effect drives people insane, save for the terminally narcissistic.* That was what it was like seeing my blog against the backdrop of the Google universe. I was data, sucked up, processed, categorized, and packaged up as needed. Google knew nothing of books or of me; it only knew the words I used and the combinations I used them in, and when to show them in search results. I was a drop in the ocean.

That constant reminder of my own insignificance stayed with me. I also had the sense of my soul being split; on the one hand, I was helping

* In fact, the character who retains his sanity, Zaphod Beeblebrox, survives not because of his egomania, but because he *is*, in fact, the most important person in the universe.

architect the machine that processed the world's data. On the other, I was one inconsequential piece of that data.

That indifference registered again with Google Books. Beginning in 2002, Google scanned around twenty million books via university libraries and their own high-speed scanning and page-turning machines. The pre-copyright books were not a problem, and they are still accessible on Google Books today. Google had hoped to strike a deal with publishers to make itself the broker for sales of existing online books that were not free of copyright, but this was not to be. The Authors Guild filed suit against Google in 2005, an ambitious settlement was rejected in 2011, and the case was finally decided in Google's favor in 2016, leaving Google free to use the books it scanned in its search results but not to sell them. Google does sell some ebooks as of this writing, but only a tiny percentage of the copyrighted works it has scanned.* In the absence of a significant financial incentive, Google Books remains a treasure trove, a vast compendium of works written before 1923, most obscure and forgotten: Victorian potboilers, antiquated reference works, chronicles of trends long past. It is also, however, a flawed archive. The character recognition is unreliable, the text formatting of the resulting e-books is irregular and sometimes unreadable (especially in the case of non-Latin characters), and I've sometimes stumbled on a scanned page that is folded in half or otherwise obscured, when the page-turning machine fell prey to wind or some other mechanical irregularity.

Google Books was, ultimately, less an organized archive than a heap of stuff. A great deal of the world's information had been gathered in Google Books, but it was irregular, flawed, and incomplete. To Google it was a dataset, no more or less perfect than any other. Perhaps if some careful algorithmic analyses had been run over the books, Google's machines could have found irregularities that could be conclusively identified as errors. But there was little opportunity for profit. Those mistakes were left to humans to discover.

Google nonetheless created the largest online library, even if it was dotted with random flaws and serendipitous accidents. Jorge Luis

* 1923 has remained the cutoff date for the ever-lengthening terms of copyright since. This cutoff coincides with the rise of films and, in particular, Mickey Mouse, created in 1928.

Borges's classic story "The Library of Babel" describes a near-infinite library in which every possible text exists, by virtue of there being a book possessing every possible arrangement of the twenty-two orthographic symbols of its alphabet.

> All that it is given to express, in all languages. Everything: the minutely detailed history of the future, the archangels' autobiographies, the faithful catalogue of the Library, thousands and thousands of false catalogues, the demonstration of the fallacy of those catalogues, the demonstration of the fallacy of the true catalogue, the Gnostic gospel of Basilides, the commentary on that gospel, the commentary on the commentary on that gospel, the true story of your death, the translation of every book in all languages, the interpolations of every book in all books.

It is impossible to find anything. No particular text can be located among the vast volumes. Finding the *right* book in the *right* language is a fool's errand when it is surrounded by millions of wrong books in unknown languages and no index exists. (Borges points out that such an index does in fact exist . . . in some other book in the library, never to be located.) Computers can generate such libraries, as pointless as they would be.

Google Books scanned and mis-scanned from the library of human meaning, and it did so imperfectly. What it obtained, instead of a Library of Babel, was an accidental assemblage: a partial, somewhat arbitrarily selected production of works.

In another Borges story, "The Lottery in Babylon," a society turns over the fate of its individuals and community to a shadowy Company that manages their fortunes through sacred, secret lotteries. The Company becomes more absolute and infects every aspect of life, until it fades into the chance operations of nature and life themselves, such that the very idea of the Company becomes tantamount to religion.

> [The Company's] silent functioning, comparable to God's, gives rise to all sorts of conjectures. One abominably insinuates that the Company has not existed for centuries and that the sacred disorder of our lives is purely hereditary, traditional. Another

> judges it eternal and teaches that it will last until the last night, when the last god annihilates the world. Another declares that the Company is omnipotent, but that it only has influence in tiny things: in a bird's call, in the shadings of rust and of dust, in the half dreams of dawn.

What Google possesses is not a Library of Babel but a Library of Babylon. Its mistakes in character recognition, page-turning, and cataloguing are random and unguided by human hands, subject to the caprices of physics and chaos, like the weather or the human body. Many predict the miracles of artificial intelligence, from perfect robot companionship to perfect economic management, but our ever-enlarging corporate networks are governed less by rationality than by a slapdash assemblage of heuristics and approximations that slouch toward the chance operations of the Lottery in Babylon. Such heaps of code work best when 75 to 80 percent success is good enough, as with search results or content filtering. When every mistake stands out, as with voice recognition or translation, these titanic, labyrinthine systems look far less impressive.* Google is a dumb god.

And often, a broken god. We are witness to the small failures of tech companies on a daily basis, whether it's a crashing phone or a mistargeted advertisement. Some failures are more malignant: Google "disappears" a page from its index, suggests a doctor's confidential patients as friends. A person is often left wondering why the failure happened, with no answers forthcoming. There is an old saying known as Hanlon's razor: never attribute to malice what can adequately be explained by stupidity. Goethe phrased the idea similarly in *The Sorrows of Young Werther:* "Misunderstandings and indolence cause more mishaps in this world than cunning and malice do." The world of software is full of misunderstanding and indolence both. But I offer a variation for the world of data: "Never attribute to programmer intent what can adequately be explained by incomprehensible complexity." We shouldn't overestimate the degree of control programmers have over the algorithms they design. Bizarre or offensive behavior is far more likely

* And yet the gains in voice recognition have been stunning, even if computers still have great difficulty identifying where one sentence ends and the next begins.

to be an accident than it is a consequence of the actual design of the algorithm.*

I'm asked sometimes why engineers at Google and Facebook are so arrogant to think that they can look at millions of people's data and not feel they are violating their privacy. Normally, data analysis is run by computer, but engineers do sometimes peek when testing, developing, and debugging. I don't think it is arrogance exactly; rather, engineers function at a priestly remove from the world. At companies like Google or Facebook, programmers engage with people's personal information in such a way that they are indifferent to its implications. When a scandal erupts like that around political marketers Cambridge Analytica and Facebook, in which mass outrage greeted the revelation that Cambridge Analytica had obtained personal data of tens of millions of Facebook users, the true revelation is that Cambridge Analytica is just one of thousands of companies that partake in Facebook's data—albeit one of the shadier ones. Such sporadic outrages belie the irreversible ease with which consumer profiling and analysis has permeated every aspect of our lives, creating a world of information that literally did not exist fifty years ago. Here, the issue is not algorithms but data collection itself. Data does not come with a "Use Only for Good" sign attached to it.

There is a paradox in the public debate around algorithms. Half of our cultural critics are saying, "Computers are inherently biased; that's why we need humans!" The other half are saying, "Humans are inherently biased; that's why we need computers!"† I have even seen a single critic make these two statements in a single essay. In May 2016, facing anonymous allegations that it was biased against conservative media sources, Facebook released the internal guidelines it provided to con-

* I once discussed the problem of algorithmic bias with an activist, who was convinced that some of the emergent racist and sexist classifications of algorithms could be anticipated in advance if software companies were to increase their diversity. Unfortunately, this view is too optimistic. While diversity is a laudable goal to pursue *in and of itself,* we can't expect a more diverse workforce to be better able to anticipate what are often wholly unpredictable outcomes.

† I owe this observation to Adam Elkus.

tractors whose job it was to evaluate and summarize world events for their Trending Topics feature. The guidelines, it turned out, were quite vanilla, with the work being more to summarize and categorize news stories than to filter them. Faced with a barrage of ill-informed negative coverage from liberal and conservative sources, Facebook laid off the employees who had been classifying stories and committed to making the feed more "algorithmic"—only for a new round of critics to say that in the absence of humans enforcing neutrality, the news algorithms were inevitably skewed by governmental and media manipulation. Facebook couldn't win. The best solution was to do nothing, or at least appear to do nothing. Facebook pulled back and made as little comment as possible. But software companies do not have the luxury (yet) of fully fading into the background like the Lottery of Babylon. They will continue to attract ire for their increasingly powerful interventions, even if none of us know what is actually going on behind their doors.

Descent from the Sky

> This fierce abridgement
> Hath to it circumstantial branches which
> Distinction should be rich in.
>
> —WILLIAM SHAKESPEARE, *Cymbeline*

In my five years at Google, I worked at increasing levels of abstraction, as many engineers tend to do. Software engineers aim to automate algorithmic processes that will run countless times on standardized sets of data that vary within known, specified parameters. For Google, such problems included crawling web pages, providing the best available results for search queries, displaying maps of user-specified locations, and running an email service for hundreds of millions of users. None of these tasks are permanently soluble. There are changing specifications, new requirements, new features to implement, or an ongoing need to improve the quality of the product.

A research-oriented computer scientist focuses primarily on the cutting edge, trying to find new methods at the very limits of what algorithms can do. Some areas of computer science, such as those dealing with compilers and operating systems, are considered more or less "solved problems" at this point. It's not that improvements are no longer possible, rather that the space of the underlying problem has been thoroughly explored. Compared to an area like computer vision or quantum computing (currently more dream than reality), there are many mature areas in computer science in which cutting-edge innovation is rare. These fields are, not coincidentally, those in which software engineering makes its trade. They were the fields that made *large-scale software engineering possible*. Microsoft, Apple, Amazon, and Google rely on solid and hyperefficient operating systems, compilers, and networking infrastructure. Artificial intelligence would be useful as well, but it is a far more daunting problem. The shape of software engineering formed itself around what was most easily achievable. "Low-hanging fruit" is what engineers call pretty much everything they accomplish rather than put off indefinitely. Google avoids trying to grasp the meaning of the web pages it retrieves. Apple revived its brand with new hardware rather than innovative software. Microsoft succeeded by slightly outperforming competitors.

A software engineer chooses which problems to take on. By the time I left Google, I had programmed for over twenty years, twelve of them professionally. The methodology of software engineering had become very familiar to me. That's not to say I obtained a serious familiarity with the entirety of Google nor with the many different types of software that Microsoft and Google made. But I knew what I liked: servers. There are many other flavors of servers I never worked on: file systems, build infrastructure, cluster management, security, monitoring, and more. I did not have expertise in them, but I did have a vague idea of the *shape* of the problems that each area contained. I felt reasonably confident that were I to enter one of these areas, I could pick up the required knowledge and be an effective contributor. And I knew that there would be a great degree of overlap with what I had done before, because the majority of software engineering is not about building something totally new, but piecing together existing pieces to make a specific hybrid that is tailored to the task at hand. The bits of genuinely new work are the reward you get for your less groundbreaking work.

There were Google engineers who were more talented than I was. They planned out epic projects and came up with brilliant designs that would revolutionize Google's projects. This elite cadre contained some of the sharpest minds I've ever known, and many of them wrote code that was stunning in its elegance and utility. I wasn't sure that I could ever be one of them. My career to that point hadn't distinguished me as one of the best of the best. I ranked in the top 10 percent of engineers at Google and I had been content at that. All the time I had spent in graduate classes and writing was time that could have been spent writing more and better code and trying to vault into that top 1 or 2 percent. I realized that I would *have* to double down on professional development if I were to avoid the encroaching sense of mundanity.

I was also distressed by the disconnect I felt between my work and reality. The god's-eye view of the world's data had numbed my relations to the world. Google, though it contained some of the most highly skilled of humanity, was, like all large software firms, committed to serving mediocrity *by definition*. By 2008, social networks were on the rise, and the average quality of content on the web was decreasing, flattened by large corporations into increasingly utilitarian and identical formats. There was still optimism in the air then. As of 2018, common consensus declares the unwashed internet to be a garbage dump of humanity's rejects. Even in 2008 there was an increasing sense that we, the engineers, were in a significant way *other* from the people who used our work. It was no longer 1995, when engineers made up a large component of the internet community. Increasingly we became spectators of our creations.

Mathematician Godfrey Harold Hardy discusses the split between the worldly and the theoretical in *A Mathematician's Apology*. For him, a brilliant mathematician, the descent from the Olympus of theoretical mathematics to the mundane real world was ignominious. For Hardy, mathematics was an eternal Platonic realm. Mathematical truths, whether they are genuine truths or not, are the most enduring discoveries of history—cross-cultural and seemingly eternal.* But Google engineers did not strive for truth the way that mathematics did. We built, only to rebuild again and again, sorting and reshaping the data

* No hard science can compete; perhaps only the fundamental philosophical questions of metaphysics (What is the nature of the cosmos?) and ethics (What should we do?) have had as long a lifetime—though answers to them have never been as forthcoming as they have with many questions of mathematics.

into momentarily useful and profitable heaps. Only researchers of the purest computer science aim at mathematics-like truths. Software engineering is closer to applied mathematics, at which Hardy looks down.

> But is not the position of an ordinary applied mathematician in some ways a little pathetic? If he wants to be useful, he must work in a humdrum way, and he cannot give full play to his fancy even when he wishes to rise to the heights. "Imaginary" universes are so much more beautiful than this stupidly constructed "real" one; and most of the finest products of an applied mathematician's fancy must be rejected, as soon as they have been created, for the brutal but sufficient reason that they do not fit the facts.

Applied mathematics and software engineering promise more dominion over this "stupidly constructed" universe than pure mathematics can, but the form that dominion takes is always dubious. For Hardy, writing in 1940, that dominion was warfare, the research efforts that marshaled the most advanced mathematics and physics of our day to create the mushroom clouds of *Trinity*. Today, technology's dominion is a combination of commercial and governmental interest. Threading the needle to "do good" is not easy, and Hardy's decision to retreat into the pure mathematical world as a refuge from the horrible realities of his age remains a stoic temptation. I did not have that choice available to me. But Hardy's statement of purpose was not wholly lost on me:

> Judged by all practical standards, the value of my mathematical life is nil; and outside mathematics it is trivial anyhow. I have just one chance of escaping a verdict of complete triviality, that I may be judged to have created something worth creating.

What I could create at Google I didn't find worth *my* creating. To feel ownership over my coded creations, I would need to work far more deeply and intelligently than I had, and that I lacked the incentive to do.

. . .

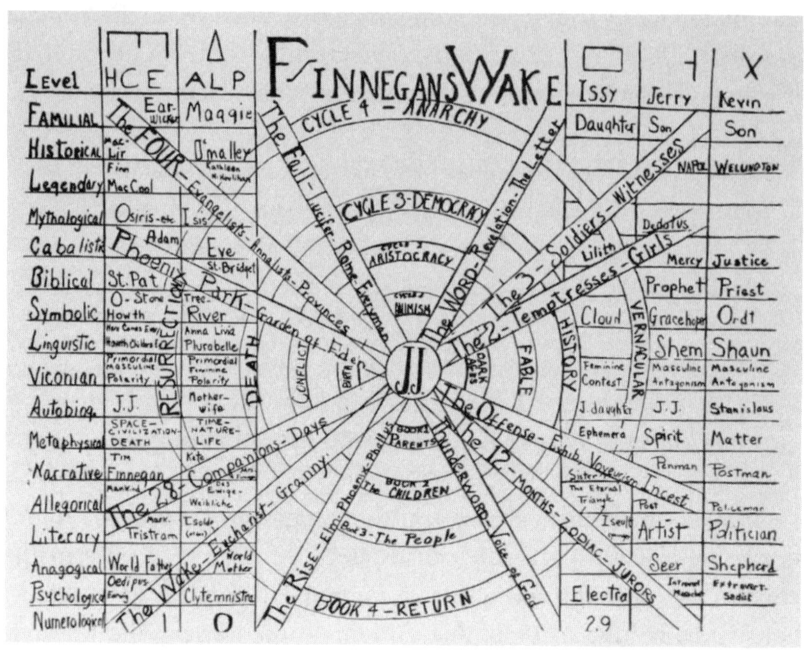

Leslie L. Lewis's infinitely contestable diagram of *Finnegans Wake*

While at Google I studied James Joyce's *Finnegans Wake* in a graduate class taught by Joyce scholar and polymath Edmund Epstein.* *Finnegans Wake* is a vast catalogue of the permutations of human existence, and it is overstuffed with meaning and structure. Interpretation on top of interpretation stack atop one another, contradictions everywhere. Difficult to parse, it was also paradoxically democratic. No one can claim to understand *Finnegans Wake* definitively. Everyone is capable of bringing something of their own to the book. Literature professor Leslie L. Lewis created a unique visual representation of the book's interlocking content and themes, superimposing the cyclical and circular over the linear and rectangular, just as the book does. Physicist Murray Gell-Mann decided that the constituent building blocks of atoms should be called "kworks," and found the phrase "Three quarks for Muster Mark" in *Finnegans Wake*. He contrived an explanation for altering its pronunciation:

* Epstein had also conducted Gilbert and Sullivan musicals (and could sing them from memory) and in 1959 had published the first bestselling edition of William Golding's *Lord of the Flies* in the United States while an editor at Capricorn Books. I was surprised to discover his name on the essay in the back of my old eighth-grade copy and realize it was the same man.

> I had been calling them quarks most of the year, but I supposed it was probably spelled k-w-o-r-k or something like that. It seemed the right sound for a new particle that was the fundamental constituent of nuclei and so on, but I didn't know how to spell it. But then paging through *Finnegans Wake,* which I had done often since my brother had brought home the first American printing in 1939, I saw "Three quarks for Muster Mark!," and of course it's "qu*ark*," it rhymes with a whole bunch of things: Mark, bark, and so on and so forth. But I wanted to pronounce it "kwork" so I invented an excuse for pronouncing it kwork—namely that Humphrey Chimpden Earwicker, whose dream is *Finnegans Wake,* the book, is a publican. And so a number of things in the book are calls for drinks at the bar. Of course, the words are multiply determined; they're portmanteau words as in *Alice [Through the Looking-Glass]*. But one determinant is often calls for drinks at the bar: "Three poss of porter pease," for example, has something to do with "Three pots of porter, please." And so here I figured that one of the contributors to "Three quarks for Muster Mark!" might be "Three quarts for Mister Mark!" And it may in fact be true.

In 1946, Adaline Glasheen, a stay-at-home mother, compiled several versions of her *Census of Finnegans Wake,* an index of all the people and characters mentioned in it. After years of the automated crunching of data with no care as to its meaning, I heard the truth of her words:

> Joyce built his house on doubt; he bet his immortal soul on the proposition that uncertainty is reality—all the reality there is; and he put his artist's money on the dark horse Incertitude. . . . *Finnegans Wake* is a model of a mysterious universe.

This would only be an academic exercise if the book were self-indulgent nonsense, but Joyce's subject was the course of human civilization and the stories we tell about it. Beneath the daunting lexical surface, the book is about the mostly tragic fates of ordinary people who suffer through the currents of history. It was the most *human* document I had read, piling on the ambiguities that my professional work demanded I simplify or ignore. Next to it my code felt meaningless.

Software engineering was tremendously pleasurable, especially when the daunting arrays of algorithms left complex trails to decipher and debug, and I enjoyed working deep in the engine room. But after twelve years, it threatened to strip meaning from the greater part of my life. *My* life, that particular set of random details picked out from Google's great library that happened to belong to me. It seemed to be nothing more than one inconsequential assemblage, a combination more or less unusual than most and no greater in significance. I was losing the ability to distinguish *my* life from the growing mass of internet data. I wanted to find myself again. Then I discovered computers had already found me.

8

PROGRAMMING MY CHILD

Initial Conditions

and errors can happen to you and computers
'cause you are . . . a computer!
go and do it!
program yourself!
just do it!
explore your toes, explore your nose,
explore everything you have goes
and if you don't want to do that
you can't even live
not even houses, 'cause houses are us

—ELEANOR AUERBACH (AGE FOUR),
"The Blah Blah Blah Song"

A FEW YEARS after leaving Google, I started another long-term engineering project, which is still ongoing. I sent this progress report to our friends after the first twelve months:

Dear Friends,

Our daughter Eleanor recently spent two weeks away with her mother visiting family and receiving upgrades. On reuniting with her, I examined her new features. Now, you know that she had had her chassis replaced with a larger one in the middle of last year, owing to the malfunction of the older model. This time there was plenty of space, but we went for the functionality upgrade.

I assume you already know that she wasn't simply changing on her own. I mean, really: you pour food in one end and suddenly she gains the ability to crawl and stand and babble? Sure, and you can grow a beanstalk up to the clouds with magic seeds. It doesn't work for computers, and it doesn't work for babies. No, in fact, all of those skills come as upgrades. Some of them can be purchased à la carte, but due to monopolies that have been long left in place due to the nefarious influence of lobbyist money on antitrust law, most of them have to be purchased as packages.

So while my daughter obtained the "standing" and "turn book pages" and "tentative stepping" and "MA MA MA" upgrade (that last one is something of a luxury item, it would seem, and in high demand), we were also stuck with her getting "throw food on the floor," "play with remote control," and "chew on glasses" (overstock, apparently). Lemon socialism is what it is. Also the enigmatic "go to sleep" feature, which has her put her head on the ground and feign sleeping, only to rise back up within seconds. Whose idea was that?

Nonetheless, on balance it certainly qualifies as an upgrade, and I fear that we filled so many of her expansion slots that we're going to be stuck upgrading the chassis again before too long. It's a shame, because she only now seems to be getting the hang of the current one. She's taken steps forward without holding onto anything and she's gained more fine motor control over her extremities, due to the extra musculo-skeletal features that we paid an arm and a leg for (actually, we paid for an arm and a leg—twice). She also seems to have figured out the rudiments of the absurdly complex paired tongue/vocal-cord action, which requires the equivalent of rubbing your belly and patting your head at the same time (which she can't do yet).

And of course incompatibility remains the norm. Our daughter is not compatible with PCMCIA cards, so we won't be able to educate her through my old laptop's eSATA drive nor have my phone connect to her over Bluetooth. Either would have been more convenient than the indirect methods we are saddled with.

Nonetheless, conditions are auspicious.

Yours in engineering,
David

Having a child turned me into a behaviorist. With age, I found it easier to model the psychological motivations of others and better predict responses to various stimuli, but adults are complex. My wife and I have shaped each other deeply, but we never *felt* that shaping happening. We noticed it when we spontaneously quoted the same *Simpsons* line, or when we became frustrated with how much explaining we had to do when speaking to anyone who *isn't* our partner.

But the stimulus-response cycle is out in the open with a child, at least initially. There is little for a baby to respond to *other* than its parents. The feedback loop created between parent and child is tight, controlled, and frequently comprehensible. I train my child to know that certain behaviors will get her fed, put to sleep, hugged, rocked, burped, and entertained. My child trains me, in turn, to respond to her cries with what she wants. You come to an accommodation; both your systems have synchronized, at least roughly, for mutual benefit (though mostly, for hers). When she is able to control her facial muscles, she learns that smiling will bring a flood of attention and positive feedback. The instinct to crawl provokes constant physical experimentation that slowly develops into crawling and then walking. The nerves and muscles get wired to successfully cope with the world. When my daughter began to crawl, I put on *Flight of the Valkyries,* because her struggle and determination reminded me of the triumph I felt on getting a particularly thorny piece of code to work correctly. She seemed delighted, like a kid with a new toy—but the toy was her own body. From birth, infants' brains are coding and debugging their own bodies.

Parents program their children—and vice versa. Psychologist Vasudevi Reddy describes the push-pull of attention this way:

> When adults at 4 months [after childbirth] start to get subconsciously desperate at infants' waning obsession with them (looking around the room, for instance, is absolutely a marker of infants

in the lab at 3 to 4 months in contrast to infants gazing exclusively at the mother's face at 2 months), they start performing more and more exaggerated actions—moving the infants' feet, singing songs, starting to invite the infant into rhythmic games, and so on to regain infant attention.

While the child's manipulation of the adult isn't quite at the level of toxoplasmosis, the parasite carried by cats which infects and alters the behavior of their owners, it's an example of how even a fairly simple relationship requires ongoing adjustments to our mental algorithms and heuristics (our own "code"). My mind rapidly changed and adjusted to being a father, turning me into a different person. My wife imagined a conversation with our one-month-old going something like this:

ME: shh, shh, go back to sleep
BABY: can't, somebody is hitting me in the face
wait why are you strapping down my arm
I need that arm to defend myself from whoever is hitting me in the face
ME: shh
BABY: you smell like milk
nghwaaaah! nghwaaaah! need milk
ME: you just had a bellyful
BABY: (screws up face) (poops) belly empty now
ME: (changes diaper)
BABY: I am pleased to be naked
let me stay here, don't make me sleep and I'll promise to not turn into an overtired human air raid siren
trust me
look at this sweet face
would I lie to you?
ME: pacifier time
BABY: no, no, no, NO, NO, NO, THIS IS AN OUTR— mmm . . .

A baby can't reason, but parents ascribe these kinds of motivations to their babies in order to explain their behavior. Sometimes they are even accurate. As our daughter grew, she learned how to express herself more precisely, eventually using words to explain what physical behavior could not.

Within days of her birth, I had already identified these facial expressions:

1. a furious glare given me when I woke her up by patting her stomach
2. intense, wide-eyed curiosity at some big blotch of color (e.g., a parent)
3. contemptuous indifference when not eating or sleeping
4. self-satisfied amusement at increasing control over her limbs
5. tranced-out bliss after eating
6. and of course the sheer joy of eating itself

I was reading into her facial expressions. But it's in part by reading into these expressions that babies learn what it is that those facial expressions *do*. The line between genetic programming and environmental programming is terminally fuzzy. So much behavior in infancy appears hard-coded, from crying to nursing to crawling to grabbing everything in sight, that I often felt like we were playing out a scripted pageant of upbringing that had been drawn up over many millennia and delivered to me through the telegrams of my DNA. But unlike many genetic burdens, such as puberty and sleep, this was one I was happy to take on, even as I reminded myself that this little creature lacked what we think of as memory and surely "experienced" things differently from mature adults.

As you age, each day, week, and month of your life takes up an increasingly smaller percentage of the time you've been alive, and so they seem to go by faster. An older man once observed to me that in terms of perceived time, life was half over by the age of twenty. If my infant daughter had any sense of time at all, each thing she did must have felt *epic* to her. A week became half her life to date! Every feeding: a monumental feast that lasts a million years! Every nap: Rip Van Winkle! Sadly, every stomachache is an aeon in purgatory and every hunger pang is a Hebraic exile in the desert, so I couldn't really blame her for crying, could I?

Sometimes I got it wrong. I once watched my infant daughter happily jiggling around and smiling as she lay on her back in her crib, around three months old, looking straight up at me. I smiled at her, then saw the clock.

"I'm sorry," I said forlornly, "but it's time to go to sleep."

Her face immediately crumpled up into an outraged grimace and she began to howl. I was momentarily amazed at her comprehension.

"That turned around fast," I said to my wife.

"You know she's responding to your tone, right?"

"Oh," I said, using my happiest chirping tone and smiling at my baby, "so if I *talk* like *this*, you'll cheer right *up*?"

Her face untwisted, she smiled and beamed at me, and she began her little lying-down dance again.

"I'm sorry," I said, keeping my tone animated and chipper, "but it's still time to go to sleep!"

She kept on bopping and smiling.

My work was done. My wife and I made a quick exit from the room, getting a few steps out the door before the surprised and indignant howls started up again, even more loudly than before. We had tricked her! But no, she wasn't likely capable of feeling *tricked* yet; her mind was bouncing between satisfied and dissatisfied states, causing her to project displays of happiness and frustration that had been hardwired into her DNA and brain. I too had been hardwired to react to them—conditioned responses that had lain dormant in my brain were on hair trigger now that there was a child to activate them. I was glad for that. In the months before her birth, my greatest fear was that I would *not* feel that visceral, emotional bond to my child, and would be able to see her only from a detached, rational perspective. I did not think that this was sufficient to make one a good parent.

I was of two minds in those early months of parenting. A part of me felt my baby was inhuman. In her first hours, she was unable to make a facial expression, unable to express anything but varying levels of comfort or discomfort. At this time, children are at their *most* transparent, as their range of responses is so limited. It's also the time in which my child seemed most like a machine. Her responses were, if not predictable, closely circumscribed. I imagined coding up a stochastic algorithm, one that relies partly on chance, to cause her to move her arms and legs jerkily, cry when hungry or uncomfortable, sleep nonstop, and nurse—not completely predictable, but rarely doing the wholly unexpected. After a few months of life, a baby has reactions that are a hundred times richer and subtler than they were at birth, and the possibility of any sort of fixed algorithm to control its behavior vanishes. I believe that part of what we consider so particular to being *human* is how we repeatedly transcend previous limitations as we grow up. To compare an infant and a toddler, or a toddler and a kindergartner, is to be amazed that one could possibly transform into the other. Each is an order of magnitude more complicated. By the time a child reaches adulthood, there have been so many overlaid levels of complexity that we must resort to abstractions and simplifications (and labels) just to get a handle on what a person is like. It is incomprehensible that this adult had been a crying baby. This process is developmental and ecological, not algorithmic.

The other part of me, though, was filled with feelings of parental love. I responded to my child's coos and cries with affection. I made faces and spoke to her and moved her feet and held her in the air like an airplane. I held her little body and rocked her back and forth, just because it made her smile. Her smile conditioned me to do it. Even the occasional variation of "airplane dumping its waste out its exhaust and onto the shirt and face below" did not dissuade me.

I played these games with her out of love and out of the knowledge that they were helpful for her development. I knew that my affection for my daughter was partial and irrational. It was the clearest evidence of psychological dispositions assigned to me by evolutionary processes. Isn't parental love one of the most powerful forces in helping babies survive to adulthood? I knew that I *needed* those irrational feelings of affection and devotion, as well as whatever chemicals and hormones they produced, to boost my spirits and help me provide the greatest amount of comfort, safety, and love to my child. So I dwelled on those feelings. I reminded myself of them. I did everything I could to maximize their presence in my mind, because I thought they were the single most important resource I could marshal for my child.

Parenting can be neurotic. Part of my brain sounded the "You just scarred her for life" alarm whenever my daughter screamed out. When she was about one year old, I took her to a music class. She played with the maracas and gently beat a drum, but then she saw the teacher, a tall bearded man with a guitar. She went into a frenzy, crying and gasping with terror and desperation. I took her out of the room and she relented, but as soon as she saw the teacher again, she was inconsolable.

I left the class early, let her flush the adrenaline out of her system, and played with her quietly for the rest of the day. I was certain the teacher had been the trigger, but I had no idea why. My wife took her to music class the next week, only to witness the exact same reaction. And for some months after that she became scared around *any* tall, swarthy, hirsute man, though none provoked her to the same degree that the music teacher had.

I was frustrated that I couldn't understand the logic behind her reactions. I knew the stimulus that triggered it. By that age, I was inured to most of her cries and screams, which were her way of expressing various desires. But during the music class, I saw real *suffering* in her eyes and heard it in her howls for the first time—not just pain or a demand, but real mental anguish. I wanted nothing more than to save her from it, to debug it out of her system. And I couldn't. All I could do was help her avoid the stimulus. Years later, my second child had an identical reaction to the same music teacher. Something was hard-coded into their genes to make sure they stayed away from tall, swarthy, bearded men with guitars.

Some stimuli are impossible to avoid. I want my children to grow up as free from the biases and preconceptions of society as possible. This is particularly true for daughters, since female social roles have historically been vastly more circumscribed and disadvantaged than male roles.

Children inherit our biases as soon as they are born. Some of these biases are matters of survival, pure and simple. Despite the moments of rebellion that are integral to growing up, most of childhood is spent absorbing one's parents' knowledge without question.

Received Ignorance

> Adjustment to objective reality does not exist for the sake of adjustment in itself. All adaptations are regulated by needs.
>
> —LEV VYGOTSKY

My heart sank the first time I went into a toy store as a parent and saw two distinct, gender-segregated aisles of toys. Dolls and dresses were on the left, cars and building blocks on the right. When I read my daughter a book about a princess rebelling against her parents for being told that she had to marry to be a proper princess, my daughter piped up indignantly and said, "Elsa [from *Frozen*] isn't married!" That's all well and good, but I would prefer to stick with Queen Alice of *Through the Looking-Glass* and Princess Ozma of the *Oz* books, who exist outside our cultural sphere altogether. The embrace of *any* dominant gender narrative, even those that seem innocuous or progressive, carries limitations with it.

I was surprised by how quickly children take up the taxonomies embedded within these narratives. By the age of three, my daughter had clear ideas of what *girls* and *boys* were like, how they differed, and how she was definitely the former. Animals, monsters, adults, dinosaurs, and pirates also all had certain characteristics that defined them. And then children add to the characteristics, as she was doing at age six:

HER: I think the rainbow should be split. Red, orange, and yellow are colors girls can wear, and blue, green, and purple are clothes boys can wear. [pointing to purple clothes] I feel sad I cannot wear these anymore.

ME: You know, I think there are no girls' colors or boys' colors.

HER: That doesn't matter.

ME: But do boys even wear purple that often?

HER: Okay, but I can't think of any better way to split it that's simple.

Psychologist Susan Gelman makes the compelling case that humans have an innate tendency toward *essentialism* that is present from early childhood on. Essentialism means that a particular entity has some underlying *essential* set of characteristics that give it its particular nature, without which it couldn't be *that thing*. Gelman found that "72% of four-year-olds and 73% of first-graders spontaneously mentioned inborn dispositions, intrinsic nature, or growth" when asked why particular animals possessed their typical properties and behaviors:

> For example, preschoolers said that a rabbit has long ears because "the egg made the [rabbit's] ears so that it had them when it hatched," or birds fly "because that's the way birds are made."

Yet in reality, there are exceptions to every "essential" characteristic. Some birds don't fly. Some rabbits don't have long ears. Some girls don't like dolls. Some boys run slower than girls. Some girls like math. Essentialist tendencies allow children to carve the world into "piecewise approximations of reality" (just as the *DSM* did) and lead them to stereotype and limit *themselves.* Even though, unlike computers, humans can accommodate contradictions and nuances, the intrinsic biases of our early taxonomies remain with us. Our received categories work *well enough* on a day-to-day basis, but in the end nearly every categorical schema is wrong, or more precisely, *incomplete* and *approximate.* But we *need* categories in order to reckon with the overwhelming amount of information we have to process. As Gelman puts it:

> Essentialism is not strictly a childhood construction. It is a framework for organizing our knowledge of the world, a framework that persists throughout life.

I wanted my daughter to be free of categorical restrictions. I wanted her to make informed choices, whether gender-typical or not. I wanted her to have *no* preconceptions of what people of a particular race were like. But my idea of a purely rational choice of her self-definition was a fantasy. Children don't start from nothing; they learn by experimenting. I had to let my daughter keep her preconceptions and hope that she would, over time, learn to see beyond them.

I avoid arguing with my daughter over her tastes. I didn't tell her she couldn't have a Barbie book she took a liking to. I didn't stop her from dressing up as Elsa. (I did tell her that the *My Little Pony* soundtrack made me want to go *Old Yeller* on Twilight Sparkle and friends.*) Instead, I tried to steer her toward more eclectic and obscure material, far from *any* mainstream. And I tried to flood her with as much material as I could to put her in mind of the multiplicity of narratives and taxonomies that existed. I read her Ellen Raskin and Daniel Pinkwater's eccentric children's books alongside *Amulet* and *Elephant and Piggie.* I showed her *Animaniacs* and *Oh! Edo Rocket* alongside *My Little Pony*

* I was, in fact, able to mount a counterattack after finding a fan-made *My Little Pony* album by "Horselover Fat" that distorted, rearranged, sped up, and otherwise mutilated two of Pinkie Pie's songs. I thought it was an improvement, but my daughter disagreed.

and *Moana*. I took her to rock climbing and ballet. I had no particular agenda other than the rejection of others' agendas. I wanted her to see that no one set of categories was trustworthy. I hoped that in seeing how all these taxonomies, schemata, and narratives were incompatible with one another, she would feel less bound to any single one. I couldn't program her, but I could perplex her.

The Child as Network

> [A child] begins to find that what these people about him say is the very best evidence of fact. So much so, that testimony is even a stronger mark of fact than the facts themselves, or rather than what must now be thought of as the appearances themselves. . . . Thus, he becomes aware of ignorance.
>
> —CHARLES SANDERS PEIRCE

It's impossible to debug a child because it's impossible to reset a life.

Programming is an iterative process. When I wrote software, I would code, test, and debug my code. After fixing a bug, I would recompile my code and start it again in its uncorrupted state, before the next bug emerged. The idea of initial conditions—the ability to restart as many times as you like—is integral to software development and to algorithms. An algorithmic recipe presumes the idea of a set of initial conditions and inputs. When an algorithm terminates, only the outputs remain. The algorithmic process itself comes to an end.* Every time an algorithm runs, it starts with virgin conditions that vary only with regard to the set of inputs. Colloquially, we can call this the *reset button*.

The scientific process depends on the reset button: the ability to conduct an experiment multiple times from identical starting conditions. In the absence of precisely identical starting conditions—whether in the study of distant stars or extremely rare circumstances or many-

* It is trivial to write a program that does not terminate, but algorithms must be finite. Fans of Gödel may be interested to know that an incompleteness proof shows that it is not possible to write a program that determines whether a given program (any program) does or does not run forever (the so-called halting problem).

varied human beings—the goal is that initial conditions are as close as possible in all relevant aspects.

But I cannot reset a human being. A child is not an algorithm. It is a persistent, evolving system. Software too is becoming a persistent system. Algorithms themselves may remain static, but they are increasingly acting on large, persistent systems that are now as important to computing as the algorithms themselves. The names of these systems include Google, Amazon, Facebook, and Twitter. These companies write software, but the products they create are *systems* or *networks*. While Microsoft had to carry over a fair amount of code from one version of Windows to the next to ensure backward compatibility, each version of Windows was a discrete program. Every time a user started up Windows, the memory of the computer was cleared and reassembled from scratch, based on the state that had been saved to disk. If Windows got into a strange state and stopped behaving well, I could reboot and, more often than not, the problem fixed itself. In the worst cases, I could reinstall Windows and have a completely fresh start.

That's not possible with systems. Constituent pieces of Google's search engine are replaced, rebooted, and subject to constant failures, but the overall *system* must be up *all the time*. There is no restarting from scratch. Google, Amazon, and Facebook are less valuable for their algorithms than for their *state:* the sum total of all the data the system contains and manipulates. None of these companies can clear out their systems and "start over," algorithmically. And neither can a child. A child starts up at birth, and her internal mechanisms produce a persistent, mutable system that is the child's body, mind, and personality. There are algorithmic, information-carrying processes that exist within our bodies, chief among them the coding, replication, modification, and transmission of DNA (and RNA), and the sheer *clarity* of our DNA operations stands in stark contrast to the messiness and apparent aimlessness of our daily lives. One of the great appeals of evolutionary psychology, the field that gave us the selfish gene and the analogous idea of a meme (a cultural idea that evolves), is that it hints that there may be a single, overriding goal driving all our biological and cultural doings: passing on our genes. Our lives are ephemeral one-shot processes, but our genes promulgate themselves to live on further. Physicist Juan Roederer has proposed that any *purposeful* activity, be it the copying of DNA, an algorithm calculating my taxes, or my telling my daughter to go to bed, relies on the system in question being able to be reset, and that the existence of such resettable systems is a defining characteristic of biological life itself.* In other words, it's the ability to start from scratch (or simulate starting over from scratch) that enables organisms to plan and execute, rather than dumbly follow the laws of nature like a rock or a planet. But even if the propagation of double helixes is the original goal of our existence, we still lack any greater understanding of the functioning of so much of the biology and culture that has built up around that goal.

In other words, what do our bodies and selves and societies do when they're *not* propagating our genes—or trying to get into a situation of propagating them? Some clues may lie in thinking about what happens to software programs when we don't shut them down and restart them, but let them linger on and evolve.

An algorithm is a finite, linear set of instructions that operates on a set of inputs to generate a particular output. *Algorithmic systems* like Google, Facebook, Amazon, and Twitter create a persistent system (or network) that modifies its behavior over time, in response to how it is used. In essence, these systems rely on *feedback:* their outputs affect the environment in which these systems exist, and the systemic environment—its users and also other algorithmic systems like it—provides new inputs that change the system further. Mathematician Norbert Wiener called such feedback systems *cybernetic,* drawing on the Greek word *kybernan* (κυβερνήν, meaning "to steer" in the sense of control, piloting, and governance). Algorithms establish and maintain these systems, but they can't predict how a system will behave at a given point in time. For that, one must know the *ongoing state* of the system.† The result is an evolving ecosystem. Programmers can code, debug, and fix algorithms, but we *acquire, train,* and *condition* a network by having algorithms operate on it. For a child, which is also a kind of evolving ecosystem, these algorithms include intrinsic biological mechanisms,

* Roederer uses the term "pragmatic information" to describe what constitutes the purpose-driven changes in a system, but since this is a distinct concept from Claude Shannon's purely statistical definition of information, which discusses only the degree of uncertain possibility in a message rather than any meaning it might have, I am mentioning "information" only in this note.

† To put it another way, for any operation, the algorithm *takes the current state of the system as one of its inputs.* New inputs do not occur in a vacuum, but are themselves dependent on past outputs produced by the system and the environment's response to those outputs.

the physical effects of its surrounding environment, and other living creatures—for example, parents who may wish they could reset their child's emotional valences on hearing a four-year-old sing this song, as I did when I asked what my daughter was sad about:

So many sad things, I can't even tell you.
They are all squished up into a ball.
Squished into a ball.
And sometimes things fall off the ball
and they go into the trash.
And I really really really love TV
and I hope I can watch it tomorrow morning.

I can't pull sadness out of my daughter's brain. So I let her watch TV, and hope it ameliorates the ball of sadness.

There are portions of a system that may be resettable—we can blank our Facebook profiles or return our immune systems to rough homeostasis—but the overall system has an ongoing, linear continuity. And indeed, we now speak of resetting aspects of the human mind, particularly when it comes to trauma and addiction. The rhetoric around Eye Movement Desensitization and Reprocessing therapy, an unorthodox method that has shown promise in treating trauma and phobias, speaks not only of desensitization but of returning the mind to equilibrium, processing and resolving a bug in the system. The system doesn't stop, nor does it return to a virgin state, but we hope that the network that makes up the human mind can be repaired on the fly.

"On the fly" is also the term used for modifying and fixing a computer program as it runs, without stopping and restarting it. While components of Google's and Facebook's networks are constantly shut down, modified, and restarted, the entire system persists and evolves. We dream of reset buttons for the soul and self, of ridding ourselves of addictions, phobias, bad habits, and the miscellaneous accumulated burdens of our lives. Now that we understand many mental illnesses to be neurochemical rather than cognitive, medicine aims to fix the apportioning of serotonin and dopamine in order to correct imbalances. People speak of the shamanic psychedelic drug ayahuasca as a sort of bleach for the mind, washing out layers of sediment that have clogged the functioning of the brain and capable of provoking imaginative powers beyond our estimates. Psychologist Benny Shanon writes, "It may very well be that [the ayahuasca experience] is the creative ability of the mind but, if so, the mind's ability to create surpasses anything we cognitive scientists ever think of." One of the appeals of MDMA, both to recreational users and to some physicians, lies in its seeming ability to turn down the amygdala's generation of negative emotion, which has shown some promise for treating social anxiety in autistic patients: "MDMA administration acutely decreases activity in the left amygdala, a brain region involved in the interpretation of negative cues, and attenuates amygdalar response and emotional reactivity to angry faces."

These treatments are not precise algorithms, but blunt hammers applied to shake up the mind's innards in the hopes of producing a desired effect. The mind is a network over which we seek to gain control, and yet we find that our ability to affect it is as clumsy and indirect as the ability of a pinball player to affect a pinball machine by pushing buttons and shoving the housing. Likewise, as computer networks grow in complexity and endure over greater lengths of time, our degree of direct control over them diminishes.

In sum, an *algorithmic system* or *algorithmic network* is a *persistent agent* produced by an *algorithm,* situated in a responsive *environment.* Once a network is in play, evolving over time and never reset to its initial state, it gains a complex existence independent of the algorithms that produced it, just as our bodies and minds gain a complex existence independent of the DNA that spawned them. These independent systems are not *coded.* Rather, they are *trained,* and they *learn.*

This means that these networks are not *fundamentally* algorithmic. Rather, they are systems that grow and evolve over time, and they are systems that *cannot* be wholly reset, for to do so would be to return the system to its starting point of ignorance and inexperience. The process of creating artificial intelligence is coming to seem less a matter of coding up algorithms and more of applying algorithms to a growing system, like pouring water on a plant—or like educating a child. Systems like Google and Facebook are the first genuine digital children.

Neural network pioneer Warren McCulloch wrote in 1951 that the distinction between machines and humans was that humans' minds reacted and adapted to their environment with the purpose of thriving in it in a multiplicity of ways:

> Why is the mind in the head? Because there, and only there, are hosts of possible connections to be formed as time and circumstance demand. Each new connection serves to set the stage for others yet to come and better fitted to adapt us to the world, for through the cortex pass the greatest inverse feedbacks whose function is the purposive life of the human intellect. The joy of creating ideals, new and eternal, in and of a world, old and temporal, robots have it not.

Not yet, anyway. McCulloch was speaking of the calculating machines of the mid-twentieth century. But now that large systems like Google and Facebook are persisting and growing for years and decades, we can contemplate the possibility of an evolving, maturing network whose intelligence is not intrinsic to its algorithms but lies in its evolved complexity, developed over great periods of time and through repeated, varied, and error-prone interactions with the world—just like a child. We don't debug these networks; we educate them.

Machine and Child Learning

> We don't need something more in order to get something more.
>
> —MURRAY GELL-MANN

I did not understand how my child was changing. She grew in size, and she used more complex sentences, but I was far less certain of what was going on inside her head. In the first months of her life, I kept a spreadsheet of her milestones, and declared a new "version" whenever my wife and I deemed her sufficiently different to appear as though a software upgrade had been installed. Hardware upgrades to her height and weight were ongoing. I had:

v1 (0 weeks): Initial conditions.
v2 (2 weeks): She smiles. Her first facial expression.
v2.1 (4 weeks): Turns head and moves arm.
v3 (9 weeks): Tracks objects with her eyes.
v3.1 (10 weeks): Sleeps through the night.
v4 (15 weeks): Grips and picks up a rattle with both hands.
v5 (22 weeks): Rolls over from back to stomach.
v5.1 (23 weeks): Wraps her arms around my neck and hugs me.

It was tempting to see these changes as upgrades because *I wasn't doing anything* to trigger them. My daughter was just figuring it out on her own! Having spent two decades of our lives in front of computers, my wife and I weren't used to seeing our "projects" alter their behavior without long and hard intervention. "Maintenance" was required (nutrition came in, waste went out), but there was no clear connection between these efforts and the changes taking place in our daughter.

"Upgrades" became more difficult to track as my daughter's skills expanded and her comprehension of the world around her developed. I settled into the rhythm of her changes, so it wasn't until eleven months in, when I had to be away from her for two weeks, that I was presented with drastic advancements in her walking and object manipulation.

Yet all these behavioral changes paled in comparison to the most mystifying and transcendent leap of all—her acquisition of language. For anyone who has wrestled with the ambiguities and frustrations of how language works (and doesn't work), seeing a child learn to maneuver its verbal gears is both revealing and confounding. My daughter learned the conventions of English through a fair amount of trial and error. As she learned more sounds and began to experiment with using words to mean more than just "I want that!" I let go of the fantasy that any sort of "upgrades" were taking place at all and I came to see her as a mysterious, ever-evolving network. There are algorithms that guide the development of the network we call the child, chief among them the workings of DNA, but those algorithms are the builders, not the building itself, and they are hidden from us.

There is a period from roughly ages two to five when children have yet to learn the full semantic logic behind words and phrases, and so their utterances have often beautiful and hilarious malapropisms. But they can also be revelatory. When, at two and a half, my daughter said, "Worms and noodles are related by long skinny things," she lumped together two entities based on superficial appearance, but she hadn't yet learned what a *relation* was. She used the word "related" in a similar way to how she'd heard it used by her parents.

These confusions can be more abstract. Children create fantasies when playing. The boundary between play and real life sometimes doesn't make sense to an adult, because children treat the play like reality while they know it isn't. Once, seeing our three-year-old playing with a stuffed turtle named Squirt, my wife asked, "Does Squirt like going to school? What did he eat for breakfast?" My daughter looked up at her quizzically and said, "Mama, do you know that Squirt is not real?"

At age three, she made up this story:

THE ADVENTURES OF LION THE LION

> There are people in the house. They are cooking food and then a camel comes in the house! The people are very confused. The people asked the camel, "What are you doing here?" But the camel didn't answer because he can't talk.

She had forgotten that the story was about a lion (named "Lion") when the time came for the main character's entrance, so she substituted a camel. She wanted a camel in the story, but she had no prepared scripts for how camels and people interact, since people talk and camels don't. So the story ended there, koan-like.

My daughter was keen to use logic to argue her position when she needed to. Sometimes it took the form of threats, particularly at bedtime: "If you don't give me any milk, I'll stay awake all night. Then you'll never get any sleep and you'll die sooner." Or her plaintive objection when she was upset and rejected our efforts to comfort her: "I want *nothing*!"

Gradually, rationality asserts itself and shoves the nonsense out of the way. By three and a half, Eleanor was modeling our motives. She didn't always do so flatteringly, as when she said to her blanket, "Now I will raspberry you. You will not like it but I enjoy it and that is why we will do it." At this point, she was able to determine that everyone around her had goals and that sometimes those goals conflicted with hers. She couldn't necessarily determine others' motivations, but she knew they were there.

Similarly, at that time, my daughter began to understand that things could be made out of other things, and sometimes there were things that she could not see. She knew human bodies were made of bones and blood, which led to this song, to the tune of "Frère Jacques":

I hear David, I hear David
Here he comes, here he comes
Now he's walking through us, now he's walking through us
Now he's wet, now he's wet

ME: Why did I get wet?
HER: You're wet with blood.
ME: Where did the blood come from?
HER: Because you walked through us.

She created the explanation out of the raw materials of her observations of the world around her and her attention to how people talked. The leap from observational data to thought is one of the most amazing and incomprehensible processes in nature. Any parent will know how baffling it is to see this happening in stages. There are limits past which a child cannot go in understanding, until one day those limits mysteriously vanish, replaced by new and deeper ones. Children come by and large to the same shared understanding that we all possess. But it can't be rushed. Even as she knew that bodies were made of blood and bones, she could not yet conceive of a hierarchy of substances, judging by the frustration she exhibited after we saw a picture of an atom in a book:

ME: *Everything* is made of atoms. Even you.
HER: No, I am made of bones.
ME: Your bones are made of atoms!
HER: No, my bones are made of . . . muscles.
ME: Your muscles are made of atoms.

HER: [very skeptical] My muscles are made of . . . muscle.
ME: Muscle is made of atoms.
HER: [Utterly fed up, turns page]

What remains a puzzle to me, and to researchers in general, is how children leap from superficial imitation and free association to reasoning. The brain grows and develops, with billions of neurons added year after year—but no matter how much memory or processing power I add to my desktop server, it never gains any new reasoning capabilities.

There are many different types of networks coming into existence besides giant informational systems like Google and Facebook. There are neural networks, deep learning networks, and belief networks, among others. All these fall under the broad rubric of *machine learning*. Today's most powerful machine learning techniques, such as those employed by Google's DeepMind, excel at recognizing similarities between explicit *patterns,* whether those patterns are made of words or pixels or sound waves. They can judge whether two passages of text have similar lexical structures and word choices, but can say little about the texts' meaning. They can determine whether a creature in a photograph looks more like a dog or a cat, but they know nothing of what a cat or a dog *is*. They can beat humans at Go, but they cannot discern whether a particular Go board is beautiful or not—unless we train an algorithm on a set of "beautiful" and "non-beautiful" boards and have it try to learn that classification. While these machine learning networks can perform feats that leave humans in the dust, they inherit contexts, standards, and judgments from humans, and they are unable to generalize from a given task to similar yet distinct tasks without human guidance. They cannot reason about the application of labels, as my daughter did at age four:

HER: These ballet shoes are so soft. I bet they are made out of polyester.
ME: Maybe they are made out of marshmallows and you could eat them.
HER: You can't eat ballet shoes.
ME: Then they aren't made out of marshmallows.
HER: Nothing's made out of marshmallows.

ME AND HER (simultaneously): Except marshmallows.

HER: Only marshmallows are made out of marshmallows. That's why they are called marshmallows. All the other names are used up by people and other stuff.

In contrast to the image classification performed by machine learning networks, children quickly learn to categorize by far more than visual similarity, and in fact learn to reject visual similarity in favor of other categorizations. As Susan Gelman found, once told a pterodactyl is a dinosaur and not a bird, even a young child will tend to infer based on that category membership *rather* than any visual similarity, in guessing, for example, that a pterodactyl does *not* live in a nest or that a dolphin *cannot* breathe underwater.

A machine learning network could not switch from visual to nonvisual categories of its own accord. It needs explicitly coded directives given by humans. It can label pictures of marshmallows as "marshmallows," but cannot infer that "marshmallows are made of marshmallows." That's meta-reasoning. But if there's any such directive encoded in our genes or in our neurons, we have yet to discover it.* I wondered how my daughter merged her cognitive skills to create a whole greater than the sum of its parts. There are competing systems in play: grammar, fantasy, rationality, narrative, taxonomy, visual imagery, and more. Somehow, they coalesce into what we think of as a rational adult. There is no clear explanation for how this occurs. Many models of development have been suggested, but the only certainty is that, somehow, rational human adults are, as far as we know, the beings most capable of coping with and exploiting the world. We are also the beings most capable of complicating the world and creating problems for ourselves, and posing existential threats to the whole planet—another by-product of our evolutionary fortunes.

The psychologist Lev Vygotsky postulated that the development of child speech and child thought were fundamentally separate, and integrated only once both had reached a certain level of maturity in toddlerhood. Until that point, speech serves a communicative but non-

* I personally am fond of Jaak Panksepp's layered models of emotions, as described in *The Archaeology of Mind: Neuroevolutionary Origins of Human Emotion.*

rational and nonrepresentational function. Then children discover that words *name* things. They aren't just noises that get the child attention or food or hugs. At that point, "thought becomes verbal, and speech rational."

Vygotsky's simple schema isn't necessarily right, but it is evocative and useful. It reminds us that a mechanistic theory of rote learning and knowledge acquisition isn't sufficient to explain how children get from point A to point grown-up. There is no single executive function running the show and directing the development of language, but multiple constituent parts that *somehow* become integrated, at which point children's reasoning power soars by leaps and bounds, and continues to do so for at least a decade. As thought becomes more abstract around the age of five, even identifying the precise changes taking place is an impossible task. My daughter, who has lost her memory of her early years as all children do, has not been of much help. At age five, I asked her about the meaning of her song, which she'd forgotten she had made up:

I hear David, I hear David
Here he comes, here he comes
Now he's walking through us, now he's walking through us
Now he's wet, now he's wet

As to why I was wet, she said, "Is it because I like to splash you with water at bathtime?" An excellent, logical explanation, which brings back many soggy memories. She was perplexed to hear the real explanation she'd given: "Did I say that?" The entire nature of the word "wet" had changed for her. It was no longer something that linked disconnected things like blood and water. It was now a real concept. Now she knew that blood had nothing to do with how *I* could get wet. In moments like this, in which the whole of child development seems packed into the changing use of a single word, I think of Vygotsky's maxim "The meaningful word is a microcosm of human consciousness."

She even came up with explanations for why she hadn't known things before. She was perplexed that there was a time when she hadn't known that the natural numbers went on forever. So she said:

> Once babies get to 100, they don't realize there are still other numbers that are hiding behind it. 100 is the king of the numbers, but it hides secrets from babies. The secrets are 101 and all of the other numbers above it.

The transition from free association to rational explanation that children unknowingly make is a mystery that artificial intelligence has yet to conquer. The problem facing AI at this time is how to move from the specific to the general in a *humanly rational* way: how to take the knowledge from one clearly defined task, like labeling images or playing Go, and put it to new and different use in a general-purpose thinking network. I suspect that accomplishing this will require the creation of networks that engage with the physical world in a variety of different ways, processing visual and verbal information in a variety of contexts and learning—slowly—what approaches do and don't work in various situations. If this were possible, the network would also need to be sufficiently powerful to apply the same broad set of techniques to varied and novel problems.* We still have a long way to go.

If we do indeed create such a general network, it's not clear that any great secret of the nature of intelligence will be revealed. We'll have created something as complicated and irreducible as a human infant itself. We will be able to watch these networks grow, learn, and mature, but we will not be able to debug them any more than we can debug a child. Nor will we understand how or why they function in the way that we understand how an algorithm functions. To say, "Oh, well, it said 'Goo' instead of 'Ga' because this set of network weights was not triggered and this one was" is not an *explanation*. Rather, we will see, as I did with my own daughter, that a complex set of predispositions and behaviors, when encoded into a single creature, results in even more complexity when that creature starts to engage with the world in myriad ways.

When Alan Turing coined the Turing Test (which, contrary to frequent news reports, has not been beaten by any computer), he implied that behavior and speech are sufficient evidence to determine whether a creature can "think." As my daughters grow up, I witness them increasingly *thinking* in ways they have never done before, just as AIs

* By "technique" I mean some low-level mechanism like a brain neuron or a hormone.

are starting to impress us with their "thinking." If something *acts* like it's thinking, that will be good enough for most people. It is no wonder we are desperate to program AIs to love us. We had better be prepared to love them as well. As my older daughter once asked me, "If we break through a screen, are we in the computer's life, and do we get to feel what it feels like?"

9

BIG HUMAN

The Vacuum Cleaner

Technology's primary effect is to amplify human forces.

—KENTARO TOYAMA, *Geek Heresy*

IF THE 2000S were the decade of Google, the 2010s have been the decade of Facebook—and social media more generally. The internet's infancy of unstructured information gave way to an adolescence of clumsy, crude social interaction. Facebook's revenues ($40 billion in 2017) remain only a third of Google's ($110 billion in 2017) and half of Microsoft's ($89 billion), less than a quarter of Amazon's ($178 billion), and all pale next to Apple's $229 billion. Yet Facebook has had a more powerful effect on *transforming* the web in the last ten years than any of those other companies. The web is becoming less centered on *pages* and more centered on *people*—more specifically, computational representations of people.

In retrospect, it was inevitable. Friendster was so hapless that it seemed like only a novelty, and MySpace was popular but ugly and underdeveloped, yet from today's vantage it seems impossible that a Facebook would *not* come along to colonize the web and the world alike. I never worked at Facebook, and so my attitude toward it is a bit like that of a land-based dinosaur observing the birds, scrawny yet with the power of flight, with a mixture of curiosity and condescension. I'm not sure that Facebook itself will outlast Google or Apple, but Facebook deployed something that will be with us for a very long time: the datafication of humans and, more significantly, our *identification* with our digital representations.

In the 1990s, anonymity ruled the web. Online celebrity was still an oxymoron. Any biographical details added to a personal web page were frequently cursory and likely to never be noticed. The early web was better suited to the spread of information than to making social connections. For many years I blogged pseudonymously, enjoying the ability to create and write in whatever voice I chose.* Many others did the same, whether anonymously, pseudonymously, or under their real—but frequently unknown—names. Beginning in 2000, the tireless culture blogger Mark Woods maintained a daily commonplace book of art and literature on his *wood s lot* blog until he passed away in 2017, revealing little of himself beyond his name. On his Frequently Asked Questions page, there was only a single blurry photo of him on the beach, and an extract from Wallace Stevens's "Peter Quince at the Clavier." For the few thousands of us who read him, he was someone we knew. We related not through personal details or life events, but by the sharing of our enthusiasms. Conversation was often more implicit than explicit. I too wrote my blog as an outpost away from the world, not as a reflection of it.

By 2010, people were beginning to trump content. Information became increasingly centralized on large sites like Wikipedia, Amazon, and Google, and *digital identity* was the new game: how to tie this diffuse, online information to identifiable individuals. Google's link graph, which mapped the influence relationships between web pages, was succeeded by Friendster, MySpace, and Facebook's friend graph, which charted the social relationships between human beings. Facebook, Twitter, and Google+ were not only social networks. They were *identity services*, attempts to bind individuals permanently to public or semipublic online identities that would be managed by corporations.

Having left Google in 2008 to return to being a digital civilian, I was

* I still blog at waggish.org. People of my generation romanticize the so-called golden days of blogging, but there's something to be said for a decentralized content publishing platform, and it is unfortunate that Facebook, Twitter, Reddit, and the like have made the act of reading content on some random website a fringe activity.

wary of these new developments. I dislike declaring any affiliation or affinity, lest I be held to it. But I seemed to be in the minority. Millions, particularly the young, signed up on MySpace, Facebook, Instagram, and elsewhere in order to publish their lives, demographics, and tastes, much as Google had classified web pages by their words and Amazon had classified products by who bought them.

The categorization and taxonomizing of human beings was not itself a new trend. Throughout the twentieth century, critics of modernity, industrialization, and capitalism, from Georg Simmel to Lewis Mumford to Jane Jacobs, had bemoaned that society was boxing people and organizing them by the work that they did. Simmel observed in his 1900 book *The Philosophy of Money* that the interchangeability of labor was generic to any large-scale economy, capitalist or not. Socialism, he suggested, required an even more generic treatment of labor, because a centrally planned economy would be less able to accommodate individual variation than a decentralized one.* The fear of automation began with the industrial revolution and accelerated with the introduction of computers, as more and more kinds of human labor began to be performed by machines. Yet computers are uniquely able to track the variations among hundreds of millions of people. This was only a distant vision, since meaningfully analyzing data on billions of people remains an extremely difficult task, but computers enabled the possibility of a digitally micromanaged society. While the industrial revolution and the advent of the assembly line generated coarse-grained classifications broken down by job requirements, the emergence of mass computation in the latter part of the twentieth century enabled large-scale, centralized classification of *individuals*.

The computerized shift toward micromanagement was driven by national defense and advertising. Advertising refined and enhanced its demographic analysis of consumer segments,† while governments initiated the trend toward computational representation and analysis

* All large industrialized economies *depend* on the interchangeability of labor, so that employment and productivity can be calculated in terms of groups of average individuals rather than collections of unique, incommensurable individuals, which would prove immune to economic analysis. Karl Marx was concerned with the *exploitation* of labor, but not the commodification of it, which is required in order to allocate work under any centralized system.

† Joseph Turow's *The Daily You* is a superb overview of the history of advertiser targeting and microtargeting.

of individuals on a mass scale. The National Security Agency, in the wake of September 11, 2001, initiated its "vacuum cleaner" approach by amassing as much data as it could in order to weed out any and all national security threats. It ended up with an ever-growing haystack containing a handful of real needles and a tremendous number of fake needles. The NSA's vacuum cleaner anticipated what would come to be called the quantified self: track *everything* in the hopes of learning *something*.

The NSA, alongside the FBI and the CIA and the TSA, assembled profiles as part of the No-Fly List, the Terrorist Screening Database, the Computer-Assisted Passenger Prescreening System, and others. It built these lists with the aid of programs like PRISM, XKeyscore, and MUSCULAR, which were designed to surveil and assemble information about specific people. This was hardly a new approach for intelligence agencies trying to identify people of interest, but these programs enabled unprecedented scope. Before the computer age, it was not feasible for organizations to compile dossiers on unlikely or irrelevant targets, because it was difficult to track such large quantities of data on paper. Storage would have been a problem, as would trying to search through the data. Consequently, the selection of what was important and what was not had to take place *before* collection. But the dawn of big data removed any practical physical limitations on storage, and amassing lots of data was far easier than *analyzing* it. At the time of the September 11 attacks, the FBI was unable to search its databases for multiple words: it could search on "flight" and "school," but not for "flight school." This crippled the utility of their data.

In *Foreign Policy*, Shane Harris described the approach of the director of the NSA under Presidents George W. Bush and Barack Obama, Keith Alexander:

> Alexander wants as much data as he can get. And he wants to hang on to it for as long as he can. To prevent the next terrorist attack, he thinks he needs to be able to see entire networks of communications and also go "back in time," as he has said publicly, to study how terrorists and their networks evolve. To find the needle in the haystack, he needs the entire haystack.
>
> "Alexander's strategy is the same as Google's: I need to get all

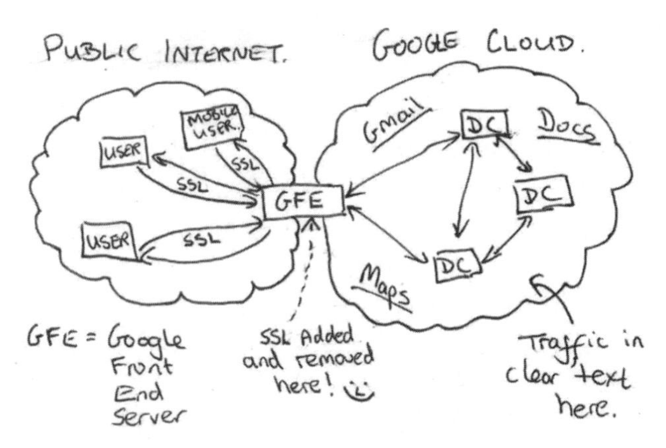

From an NSA document on the MUSCULAR program. According to the *Washington Post,* "Two engineers with close ties to Google exploded in profanity when they saw the drawing."

of the data," says a former administration official who worked with the general. "If he becomes the repository for all that data, he thinks the resources and authorities will follow."

Alexander was wrong. Having the data was not enough. A 2010 *Washington Post* exposé, "Top Secret America," revealed just how unprepared the NSA was to do the much harder job of analyzing their petabytes of data for the right signals. Half-baked tools like the "automatic ingestion manager" designed by NSA advisor and "mad scientist" James Heath were ill equipped to figure out the who and the what of the data, much less the how and why. Later documents leaked by Edward Snowden showed that NSA analysts actually begged the agency to stop collecting so much useless data. A 2010 UK report on MI5's digital intelligence capabilities concluded the same thing: there was too much data and too little analysis.

The Obama administration continued the trend of its predecessor, endorsing Alexander's approach, amassing even more data while still remaining unable to process it all. James Bamford wrote in *Foreign Policy,* summing up the Obama years:

> Alexander asked, "Why can't we collect all the signals all the time?" He applied this approach in Iraq, pulling intelligence from

> phone interceptions, planes, drones, satellites, and other sensors into a powerful computer analysis system known as the Real Time Regional Gateway. He also ran the NSA's massive metadata surveillance program, which involved secretly keeping track of every phone in the United States: what numbers were called, from where, and exactly when—billions of communications each year. . . .
>
> Privacy hasn't been traded for security, but for the government hoarding more data than it knows how to handle. Kinne, the former intercept operator, described her work as "just like searching blindly through all these cuts to see what the hell was what."

Tracking "every phone in the United States" was impossible before the era of big data. In the twenty-first century, it was not just possible, but aggressively embraced. Companies like Facebook and Google still need users to bring their data to them, to a point. But government organizations show us where corporations are headed, as data consolidation continues apace.

Once data is collected, there are no intrinsic restrictions on its use. The original purpose of this collection is to intercept communications and prevent terror, but because the data dragnet is so total, other uses opened up.

> In a top-secret memo dated Oct. 3, 2012, Alexander raised the possibility of using vulnerabilities discovered in mass data—"viewing sexually explicit material online," for instance—to damage reputations. The agency could, say, smear individuals it believed were radicalizing others in an effort to diminish their influence.

When government agencies—or corporations, for that matter—combine this aggressive anti-privacy stance with the inevitable mistakes profiling systems make, the potential for chaotic and ubiquitous abuse balloons.

Profiles

> I was left with this surrogate mirror
> I thought: Who created this monster?
>
> —THE FALL, "Surrogate Mirage"

In order to identify terror suspects, the NSA needed to classify everybody as a likely terrorist or an unlikely terrorist. In other words, they applied labels. Data collection and storage have become cheap, and the vacuum cleaner approach has been adopted by private corporations as well. Consumer profiling has long been a staple of marketing data providers like Experian and Acxiom, but Facebook has become a centralization point for the *collection* of personal information in order to target individual consumers. Facebook has sorted its users into a large number of categories and buckets, assigning them advertiser-friendly demographic labels. Here are some of the axes on which Facebook allows advertisers, data analysts, and other third parties to "microtarget" users:

1. Location
2. Age
3. Generation
4. Gender
5. Language
6. Education level
7. Field of study
8. School
9. Ethnic affinity
10. Income and net worth
11. Home ownership and type
12. Home value
13. Property size
14. Square footage of home
15. Year home was built
16. Household composition
17. Users who have an anniversary within thirty days

18. Users who are away from family or hometown
19. Users who are friends with someone who has an anniversary, is newly married or engaged, recently moved, or has an upcoming birthday
20. Users in long-distance relationships
21. Users in new relationships
22. Users who have new jobs
23. Users who are newly engaged
24. Users who are newly married
25. Users who have recently moved
26. Users who have birthdays soon
27. Parents
28. Expectant parents
29. Mothers, divided by "type" (soccer, trendy, etc.)
30. Users who are likely to engage in politics
31. Conservatives and liberals
32. Relationship status
33. Employer
34. Industry
35. Job title
36. Office type
37. Interests
38. Users who own motorcycles
39. Users who plan to buy a car (and what kind/brand of car, and how soon)
40. Users who bought auto parts or accessories recently
41. Users who are likely to need auto parts or services
42. Style and brand of car you drive
43. Year car was bought
44. Age of car
45. How much money user is likely to spend on next car
46. Where user is likely to buy next car
47. How many employees your company has
48. Users who own small businesses
49. Users who work in management or are executives
50. Users who have donated to charity (divided by type)
51. Operating system

52. Users who play canvas games
53. Users who own a gaming console
54. Users who have created a Facebook event
55. Users who have used Facebook Payments
56. Users who have spent more than average on Facebook Payments
57. Users who administer a Facebook page
58. Users who have recently uploaded photos to Facebook
59. Internet browser
60. Email service
61. Early/late adopters of technology
62. Expats (divided by what country they are from originally)
63. Users who belong to a credit union, national bank, or regional bank
64. Users who invest (divided by investment type)
65. Number of credit lines
66. Users who are active credit card users
67. Credit card type
68. Users who have a debit card
69. Users who carry a balance on their credit card
70. Users who listen to the radio
71. Preference in TV shows
72. Users who use a mobile device (divided by what brand they use)
73. Internet connection type
74. Users who recently acquired a smartphone or tablet
75. Users who access the internet through a smartphone or tablet
76. Users who use coupons
77. Types of clothing user's household buys
78. Time of year user's household shops most
79. Users who are "heavy" buyers of beer, wine, or spirits
80. Users who buy groceries (and what kinds)
81. Users who buy beauty products
82. Users who buy allergy medications, cough/cold medications, pain relief products, and over-the-counter meds
83. Users who spend money on household products

84. Users who spend money on products for kids or pets, and what kinds of pets
85. Users whose household makes more purchases than is average
86. Users who tend to shop online (or off)
87. Types of restaurants user eats at
88. Kinds of stores user shops at
89. Users who are "receptive" to offers from companies offering online auto insurance, higher education, or mortgages, and prepaid debit cards/satellite TV
90. Length of time user has lived in house
91. Users who are likely to move soon
92. Users who are interested in the Olympics, fall football, cricket, or Ramadan
93. Users who travel frequently, for work or pleasure
94. Users who commute to work
95. Types of vacations user tends to go on
96. Users who recently returned from a trip
97. Users who recently used a travel app
98. Users who participate in a timeshare

Much of this data is not directly provided to Facebook by users. Facebook gathers information on us even when we aren't explicitly providing it. In addition to what it collects from profiles, pictures, and clicks, Facebook correlates its information with what it obtains from third-party sources—such as car registrations, residential information, and corporate information.

For each of its users (as well as for many people who *don't* use Facebook), Facebook creates a detailed shadow, and it profits by presenting these shadows as valuable targets for the marketing of goods, services, and ideas. Facebook's marketing partners, whether local businesses or scam artists or shady political operatives, use this data not just for marketing but for their own surveillance of human behavior. Taken together, the politics, habits, demographics, hobbies, and personal relationships of a Facebook user allow for a greater degree of persuasion—and potential *control*—than ever before. Health information is protected by law, for example, but given a person's drugstore purchases, alcohol habits, and life events, any company could make some very educated guesses

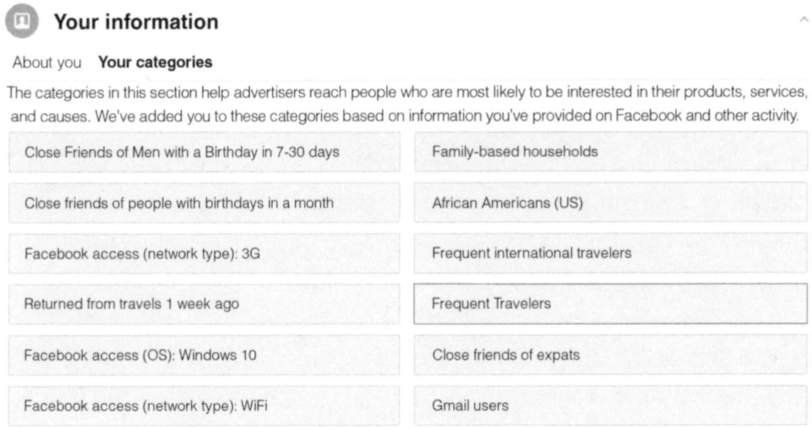

Your information

About you **Your categories**

The categories in this section help advertisers reach people who are most likely to be interested in their products, services, and causes. We've added you to these categories based on information you've provided on Facebook and other activity.

Close Friends of Men with a Birthday in 7-30 days	Family-based households
Close friends of people with birthdays in a month	African Americans (US)
Facebook access (network type): 3G	Frequent international travelers
Returned from travels 1 week ago	Frequent Travelers
Facebook access (OS): Windows 10	Close friends of expats
Facebook access (network type): WiFi	Gmail users

The advertiser categories Facebook has assigned to me.
Some are correct; some are not.

as to the state of their health. Collect enough public information, shake it vigorously, and private information will fall out of it.

Facebook silently classifies users by "Ethnic Affinity" in order to target ads, a category that it infers from the entirety of a user's profile—tastes, friends, location, habits, etc. Such categories can be wrong, and most users will never even see the mistake. Last year, Facebook told me my ethnic affinity was Asian. This year, they think I'm African American (or, at least, someone "whose activity on Facebook aligns with African American multicultural affinity"). And even if you don't mark your political orientation (a spectrum from "Very Conservative" to "Very Liberal"), Facebook makes a guess at that too.

To Facebook, targeting is a matter of profit and loss rather than life and death. Yet its behavior has an enormous impact on our lives. Facebook's ethnic affinities cover only a handful of the countless ethnicities in the world—the only options are Asian American, African American, and Hispanic—yet it's *Facebook*'s categories by which you are identified to advertisers worldwide.* Products have been targeted toward particular ethnicities for decades. But it's one thing to appeal broadly

* In July 2017, Facebook quietly renamed "Ethnic Affinity" as "Multicultural Affinity," without changing the nature of the category at all. Both are euphemisms for race.

to a demographic, and another to target individual consumers by race, which means classifying each individual as belonging to one "ethnic affinity" or another—let's just call it race. As we see with other taxonomies, it is very hard to make such a classification without smuggling in all sorts of assumptions and biases. Here is one way to encode racial categories, following the taxonomy of the 2010 U.S. Census:

```
enum Race {
  White = 0,
  Black = 1,
  Native American = 2,
  Asian = 3,
  Pacific Islander = 4,
  Other = 5,
};
```

This is controversial territory. The reasons for choosing these five "races" aren't apparent and owe as much to historical accident as to any purported biology. For comparison, here is an encoding for the 1939 classification created by respected American anthropologist Carleton S. Coon:

```
enum Race {
  Caucasoid = 0,
  Mongoloid = 1,
  Negroid = 2,
  Capoid = 3,
  Australoid = 4,
};
```

It's just a number to a computer, but a wholly different classification to us. Even if there is some scientific concept that deserves to be termed "race" (a question that seems, at the least, unresolved), it does not match up with any degree of rigor to anything that is popularly called "race"—and our shifting, amorphous notion of "race" is a social construct. It functions in much the same way as Myers-Briggs or the *DSM*, except with far more reaching consequences and far less logic: the census offers no options for those with two parents of different "races" other than a nebulous "Other" checkbox. These classifications, then, are akin to what *DSM* psychiatrist Allen Frances referred to as "temporarily useful diagnostic constructs."* With race, however, the labels stick earlier and more permanently. We are slotted into categories before we are born.

To a computer, this data is neutral: the first set of racial labels I gave above are represented as the numbers 0 through 5, the second as 0 through 4. Computers are ignorant of the moiré of shifting meanings imposed on the labels assigned to those numbers, as well as whether those meanings are at all just.

Such a classification becomes an ontology to us, a way of seeing and carving up the world. Computers chronically reinforce the ontologies fed into them. When we select "Black" or "White" on a form, whether on the census or on Facebook, we render the underlying taxonomy real. We turn it into an ontology. When Facebook intuits a person's race from his or her interests and posts, they are stereotyping—in my case, incorrectly. Since these categories are as prescriptive as they are descriptive, labels and associations are reinforced without any particular consideration as to *what* is being reinforced.

In the case of the racial categories of the U.S. Census listed above, what would my own "mixed-race" children select? There's the catchall "Other," but that merely signifies that the categories need to be revised to be more comprehensive and precise. But computers don't deal well with shifting ontologies. Once they have an ontology, computers reify it. If we classify people in black and white, we bias ourselves to ignore all the factors that get lost in the cracks.

On Facebook, gender is one of the primary factors that determine what kinds of ads we see. Despite Facebook's seeming embrace of gender multiplicity (57 in the United States, 71 in the UK), advertisers choose from only three options: male, female, and all. As a consequence, men and women have *very* different experiences on Facebook. Women will see ads for beauty products, home supplies, feminine hygiene, and other products targeted at female-dominant demographics. Men will see ads for technology, sports, and cars. This *is* bias. Facebook's algorithms

* For example, "Hispanic" is not a race, yet we treat it like one. The term is more or less meaningless outside the United States, where it performs a specific political function grouping a wide swath of people together who could never fall under a single "racial" category—except as a marker of nonwhite identity.

classify ads and people by the numbers given to them. The meanings of these classifications emerge *implicitly* from what data gets shunted under one label or the other. The effect is that men and women are encouraged to keep consuming along a firm divide. Crossover is discouraged. It's hard to gauge the severity of impact of such segregation,* but I'm telling my daughter to use an ad-blocker.

Language, and our use of language, inevitably carry bias. Computer code itself lacks any such bias. But the computer data our software processes reflects life, and so it reflects our blind spots and prejudices. Once a computer starts to speak the language of humans and human practices, it plays out our linguistic biases. To be labeled is to be prejudged. By standardizing classification and making explicit our social classifications, computers have amplified the gaps and biases in our concepts to their breaking point.

Bad Labels

> I have spent much of my life turning away from the scripts given to me, in China and in America; my refusal to be defined by the will of others is my one and only political statement.
>
> —YIYUN LI,
> *Dear Friend, from My Life I Write to You in Your Life*

Facebook's targeting categories are cases of labels being applied to data objects—in Facebook's instance, people. What happens, however, when a computer tries to label an object without knowing what it is?

In 2015, Google rolled out a new feature called photo categorization, which sorts users' photos into folders based on their subject matter—what's pictured in the photos. Photo categorization assigned labels to photos, then sorted the photos by label. With this program, Google stumbled into a minefield. I heard about the first issue when my wife

* Part of this segregation owes to the reinforcement mechanisms of consumer society itself, which are merely being amplified by targeted advertising. But targeting introduces an emergent element of bias.

sent me a link with the email subject "very bad machine learning false positive." The link took me to tweets by a Haitian American software developer, who took a series of photos of himself with a female friend, and found that Google Photos had placed them into a folder tagged "gorillas," because Google's machine learning algorithms decided—wrongly—that there were gorillas in the photos.

Image recognition is an inexact science, based on "training" persistent machine learning networks through feedback mechanisms. Google's systems made a horrendous error because the label assignment invoked a legacy of racism and dehumanization. Google apologized and quickly rolled out a fix, but didn't explain what had happened and didn't assuage concerns about the possibility of the error happening again.

There were three problems. First, the word "gorillas" was not being regulated as a potentially sensitive or offensive word. Google had surely dealt with sensitivities around that word at some point in the past, and so this error seems genuinely sloppy. They quickly restricted use of the word, which put a stop to the problem. But there likely still remain other loaded words that are similarly unrestricted.

The second problem was the miscategorization of the friends' faces as animals. A higher-order algorithm failed to recognize that the photo contained *human faces*. Facial recognition algorithms such as those used by the FBI are frequently distinct from general object recognition algorithms, as faces are distinguished by variations quite specific to human physiology. But before a facial recognition subsystem can kick in, the general image recognition algorithm must determine whether a particular photo contains a human face.

Google trained its image recognition networks on hundreds of millions of faces; what caused it to make a false negative? It could well have been the subjects' skin color. Color plays a large part in what is termed "skin detection," which in turn plays a large part in face detection. (Image search engines use this approach to detect porn: safety filters that restrict image results to work-safe images determine which pictures show human skin, and then whether that skin belongs to certain body parts.) "Marginal" cases, or people with skin tones deemed too far from the overall average, are most likely to generate wrong results. It's a problem when those marginal cases happen to be subjects who are also culturally marginalized by race. An Uber fatality when one of its self-driving cars struck a pedestrian in Arizona made clear the life-or-death stakes of computer vision algorithms. Self-driving cars must avoid humans even at the cost of hitting another object or an animal. But if a self-driving car must avoid *two* humans, and only one of them registers to the car as a human, the consequences could be disastrous. And if an error comes about due to a failure to recognize humanness based on skin color, then that algorithm would deserve to be called racist.

The final problem is the intrinsic difficulty of labeling images in the absence of cultural and semantic context. Also in 2015, Flickr launched a similar auto-tagging feature. One user posted a photo of the gates of the Dachau concentration camp, only to be given the suggestions "Dachau" *and* "jungle gym." How is Flickr to detect such a distasteful (if not exactly racist) association? People can be hired to identify specific images as culturally sensitive. But for computers to do this, they would need cross-cultural knowledge beyond the possession of any single person or community. Humans have a great talent for drawing implicit associations. We also have a great talent for making some of those associations unpleasant and offensive. But there is no repository of that knowledge from which computers can learn. So computers move naively, Candide-like, and make missteps that only we can correct.

Compounding this problem is the fact that there often *are* bad actors who inject prejudice and bigotry into rich data. Google faced a much bigger problem when its Maps product started labeling geographical locations with some terms used on the web to describe those locations. As Google's Jen Fitzpatrick put it, "Certain offensive search terms were triggering unexpected maps results, typically because people had used the offensive term in online discussions of the place." In practice, this equated to Obama-era searches for "n—ga house" returning the White House and searches for "n—r university" returning Howard University. What had happened was that enough users had, in discussing the White House and Howard University, referred to them in such ways, and in its sweep of the web, Google's knowledge engine had swept up the offensive labels and found them to be common enough to associate with those locations. While Google filtered out these terms to prevent them from showing up in the info boxes in the map results, Google

Maps neglected to prevent searches *on* the problematic terms from returning offensive results.

In a subtler case, a 2016 paper by two electrical engineering professors specializing in image processing claimed to be able to group faces as "criminal" or "non-criminal" based on a machine learning analysis of facial pictures. Analyzing 1,126 internet photos of Chinese men and 730 photos of Chinese criminal males obtained from the Chinese government, they made assertions about the link of physiognomy to criminal behavior such as this:

> the angle θ from nose tip to two mouth corners is on average 19.6% smaller for criminals than for non-criminals and has a larger variance. Also, the upper lip curvature ρ is on average 23.4% larger for criminals than for non-criminals. On the other hand, the distance d between two eye inner corners for criminals is slightly narrower (5.6%) than for non-criminals.

This analysis assumed that the photographic subjects all held the same neutral facial expression, but criminal mug shots are likely to be taken under conditions that produce rather different facial expressions than ordinary photographs. And the physiological differences, particularly those final 5.6 percentage points, are not broken down enough in the paper to establish whether they are statistically significant.* The authors concluded:

> Although criminals are a small minority in total population, they have appreciably greater variations in facial appearance than general public. This coincides with the fact that all law-biding [sic] citizens share many common social attributes, whereas criminals tend to have very different characteristics and circumstances, some of which are quite unique of the individual's own.

There are no grounds for the "fact" that criminals' personalities vary more than noncriminals', nor that this correlates to a greater variance in facial features. It's not clear—and seems unlikely—that the criminals

* A substantial portion of social science research has been found to be unreproducible—over half of all studies, according to a 2015 study—and my first reaction to any bold claim is now skepticism.

were sampled from the same population in the same manner as the noncriminals. Finally, what of the noncriminals? Are internet photos a truly representative sampling of people's faces, or are they more likely to skew toward humans' preferences for more beautiful and attractive faces, or faces shot in ways that make them *seem* more beautiful? In any event, it's reasonable to expect a divergence between a set of mug shots and a set of photos culled from the internet, and it's also reasonable to doubt that this divergence is reflective of any *real* physiological difference. The real classification happened at the onset of the study, with the division into criminal and noncriminal. The analysis begged the question.

Anthropologist and eugenicist Francis Galton's classification of facial types of criminals and noncriminals, 1879

In this research, machine learning served to exaggerate certain feature differences and manufacture a classification. The mystique of data and "impartial" computation gives scientific veneer to this modern phrenology. The researchers boast that they use computers to neutralize human bias:

> Relatively little study has been done on the accuracy of character inference based solely on still face images. This is probably due to, aside from the historical controversies surrounding the inquiry and stigmas associated with social Darwinism, the dif-

> ficulty to neutralize all possible prejudice and preconditioning of human experimenters and subjects when assessing the accuracy of face-induced inference on socially charged matters such as criminality. In this work, we adopt the approach of data-driven machine learning to fully automate the assessment process, and purposefully take any subtle human factors out of the assessment process.

They have it backward. "Data laundering," where human biases and predispositions are fed into algorithms in order to make them look "objective," doesn't remove those "subtle human factors." Rather, it disguises and then amplifies them into harmful, binding taxonomies. This application of machine learning threatens to predetermine whether people are criminal simply based on the requirement to perform such a classification, whether differentiating factors are found or not.

Thus, people are stigmatized by computers. Sociologist Erving Goffman describes the function of stigma as follows:

> By definition, of course, we believe the person with a stigma is not quite human. On this assumption we exercise varieties of discrimination, through which we effectively, if often unthinkingly, reduce his life chances. We construct a stigma theory, an ideology to explain his inferiority and account for the danger he represents, sometimes rationalizing an animosity based on other differences, such as those of social class. We use specific stigma terms such as cripple, bastard, moron in our daily discourse as a source of metaphor and imagery, typically without giving thought to the original meaning.

Stigma is a form of bias: it's socially justified bias. Computers make our stigmas more explicit and more public by providing "evidence" for a stigma assignment. They thoughtlessly apply labels without regard to the biases that may be attached to them. When these labels are derogatory—think "obese," "criminal," "poor," "uneducated," and the like—stigmas are empowered and circulated by the spread of these labels through social systems.

It is to our credit that society has gradually fought against the wounds of stigmas. The stigmas of being poor or mentally ill are no longer as dehumanizing as they once were, even though they still carry negative associations. Yet our ability to retract stigmas relies on our being able to alter labels, to forget them or forget their associations. Computers, in contrast, cannot forget. They fix and perpetuate the labels they apply. Fifty years ago, society treated homosexuality as a mental disorder. In another fifty years, there will surely be biases we endorse today that will be condemned as backward, narrow-minded, and reactionary.

Algorithms inevitably make many mistakes in tracking our biases and reflecting them back to us in a sufficiently sanitized form. Keeping computers from making such mistakes is a Sisyphean task, and companies like Google and Facebook will continue to apologize whenever an offensive error occurs. This is a problem that can only be managed, not solved. At worst, computers will encourage us to keep applying certain biases even if we wish to stop.

In my time as an engineer, I was fortunate not to have to traffic in the task of having computers label people. I was not involved in distinguishing between the offensive and inoffensive. Shortly after September 11, 2001, there was a curious incident at Microsoft in which they were accused of hiding anti-Semitic code in the Wingdings symbol font. In Wingdings, the letters "NYC" generated these three symbols:

It was, in fact, a complete coincidence, though conspiracy theorists and the *New York Post* accused Microsoft of anti-Semitism. From this incident I realized that as data and networks expand, offense will sooner or later be guaranteed, no matter how smart the algorithm. We might work hard to curtail biased application of labels to data, but when the offense arises from the sheer multiplicity of labels and data, the problem is endemic to labeling itself. To label data is to inject bias into it.

As a child, I had been drawn to computers because they were free of society's tortuous value systems. Ironically, I now live in a world where computers are the thoughtless arbiters of those very same value systems. They have come to speak our languages like idiots, replicating their stigmas and biases. This is the fault of our dependence on inexact ontologies and their labels. Whether humans or computers are applying the bias, one does not eliminate bias from an ontology—one only copes with it. Our social networks today encourage us to prejudge one another, because on them we encounter one another by our labels and our statistics.

The Social Graph

> I could escape the reduction imposed by unjust laws and customs, but not that imposed by ideas which defined me as no more than the sum of those laws and customs.
>
> —RALPH ELLISON, "The World and the Jug"

Friendster, Myspace, and then Facebook and LinkedIn invented the social web as we know it. The social web is a network of friends. Relationships tend to be mutual and symmetrical (though not always, particularly on Twitter where one can "Follow" celebrities who don't follow back). Each person or user is a node in the network, and the links between the nodes reflect friendships.

This initial model of a friends network has lost significance as the social web has grown. Facebook and the social web have moved far beyond treating humans and their relationships as a network of personal relationships.

Unlike web pages, humans possess vast quantities of *metadata*. Google was able to draw a tremendous amount of information from what was primarily an unstructured (or barely structured) mass of text, which frequently gave little to no clue about its underlying meaning. Google scanned web pages for individual words. The words that were less common often functioned as coded *labels* indicating what sort of a web page it was. But a human is not a page of text. Humans are not made up of words. From the perspective of a cloud service, humans generate status updates in the form of text and photos, as well as exchanging text and image messages with one another, but these were far less useful for determining people's essential nature, or even for determining their age, gender, and race. Facebook was the first software company to confront the question of *how to determine what a person is about.* In other words, *how to label a person.*

This is a profound and fundamental question. What the cloud *knows* about us is what it can algorithmically assess, and what it can algorithmically assess determines how the cloud responds to us and recommends to us. The cloud's judgments are not only *descriptions* of the sort of people it thinks we are and assumes that we will continue to be, but also *prescriptions* for how we are to be treated. Computers classify and describe us, and we respond to their classifications, which in turn cause computers to refine their descriptions. Our behavior shapes computers, and computers' behavior shapes us.

Yet how does this feedback process take place? Social networks treat people not like web pages, but like products. A product on Amazon has a description, reviews, barcodes, related products, and such associated with it, but it doesn't reside on a single website or an authoritative single page. So it is with humans. Information, scattered across different documents, describes a person who exists outside the web. Across the web, there are hundreds if not thousands of data sources about each person. If I were to collate all this information, gather the documents concerning a particular "human" in a single place, and analyze them, I could learn a great deal about them. I'm stingy with my personal details on Facebook, but if you correlated my Facebook profile with my Amazon purchases, my websites, my public writings, my address and demographics, my credit ratings, and other such data, you could put together a detailed picture of my life, my habits, and my tastes.

Imagine, instead of a social network where each person is a page, that each person is a website containing thousands and even millions of pages. Each page holds some bit of online information from a person: status updates, tweets, photos, emails, documents, demographic data, consumer purchases, work history, education, medical history, and more. No single service will have access to *all* of a person's information, but each service will lump together what it *can* find for each individual and analyze away. Some bits of information are not especially meaningful. What can we learn about a person from a twenty-word status update like "I had a lousy day today and I'm going for a drink"? "Drink" is the most useful word, and even it is quite vague without greater context. We can't even safely conclude that it means alcohol. But if the cloud knew what the person liked to drink and how often, then it could target more persuasive advertisements to this person. If a person posted about "the Second Amendment," it could show him or her ads for the NRA, or for NRA-friendly candidates. To perform such feats, the cloud needs to accumulate metadata about a person—or better yet, have the person provide it.

The ubiquitous "Like" button, which Facebook has seeded across the web, was the simplest and most powerful generator of human metadata yet seen. By "Like"-ing a status, a page, or a product, we tell Facebook that we are *interested* in something. Deriving interests from status updates is a dodgy business; just figuring out whether something's being mentioned positively or negatively (such as the Second Amendment) is hard enough.* Photos are tricky too; just what is that person drinking in that photo—and wait, who even is that person? Image analysis was quite primitive when Facebook started out, and text can be ambiguous. But a "Like" is simple and clear-cut. It is a low-cost and high-information tool for assembling an ad hoc but very powerful profile on a user, in order to determine *what a person is about.* By federating the "Like" button across other websites, Facebook followed its users beyond the confines of Facebook proper and refined its user profiles. Facebook's idea of a "person" became a powerful, if messy, profile, full of rich metadata on a person's likes and dislikes, habits, and relationships, and a great deal of their biography. This was *useful* metadata.

Every time a person clicks the "Like" button for X, whether X is a product or an artist or a store or an idea, Facebook labels that person as someone who likes X. This was already a step further than what Facebook could get from most status updates or photos. From there, it's simple to group together people who like X, *even if they have no explicit connection to one another.* Facebook can surmise things about these groups, to infer other shared preferences. This was an application of Amazon's recommendation engine on a far greater scale. Amazon recommends products based on what people with similar purchase histories have bought, just as Netflix does with film recommendations, but Facebook knows not only what we buy, but also what television shows we watch, food and drink we consume, sports teams we root for, stores we visit, cars we want but can't afford to buy, politicians we

* The field of "sentiment analysis," which tries to gauge the emotion, positive or negative, behind a particular piece of text or a mention of a particular person or object, is plagued with problems, such as thinking "I am not happy" is a positive and optimistic sentiment because it mentions the word "happy."

support or hate, and much more. Amazon used their technology to *sell* stuff and generate enormous revenue, but Facebook was more capable than Amazon of putting together comprehensive user profiles. They discovered, after Google and Amazon, a third way of organizing the web. Google organized it by pages, Amazon by products, and Facebook by people.

A "Like" does not have to be explicit. As long as we are logged in to a Facebook account, Facebook can generate meaningful metadata about us by tracking which links we click on and which pages we visit. Facebook assembles a second set of implicit interests by tracking users' activity on and off Facebook.* This in turn assigns further labels to users: *Times* readers, car lovers, gamblers, alcoholics, etc. Quizzes are also useful: introverts or extroverts, Snapes or Dumbledores, masterminds or crackpots. Basic demographic data is increasingly available to any large corporate entity through either explicit declaration or implicit inference: gender, race, location, income class, social class, familial relationships, education history, and work history. Taken together, this data enables Facebook to profile a person in far greater detail than any number of status updates alone could allow.

Labels beget more labels. People who Liked Amstel and Corona were almost certain to Like beer, even if they never bothered to click Like on the generic beer page. If you Liked the retro indie stylings of Modest Mouse and Interpol, you were a likely bet for the Strokes, as well as indie rock communities. If you Liked Boston-area restaurants, perhaps you would like other Boston-area restaurants, especially those that paid for advertising on Facebook. If you liked InfoWars, you were a good bet for other conspiracy-theorist websites. The ever-present Suggestion box was there to suggest more pages to Like.

Social networks and their third-party partners have proven to be brilliant collection mechanisms for understanding people in a conveniently computational way. Yet the side effects of this process are causing us to see *each other* in a more computational way.

* Facebook backed off tracking efforts like its 2007 cross-site tracking program, Beacon, as they generated negative press, but subsequently redeployed similar efforts in subtler and more extensive ways, resulting in the Everest of personal information made use of by everyone from lawyers to drug companies to political consultants.

The Presentation of Self in Internet Life

> Since you are tongue-tied and so loath to speak,
> In dumb significants proclaim your thoughts.
>
> —SHAKESPEARE, *Henry VI, Part 1*

> I feel so bad for the millennials. God, they just had their universe handed to them in hashtags.
>
> —OTTESSA MOSHFEGH

The primitive level of user feedback encouraged by online services is a feature, not a bug. It is vastly easier for a computer to make sense out of a "Like" or a "★★★★★" than to parse meaning out of raw text. Yelp's user reviews are a necessary part of their appeal to restaurant-goers, but Yelp could not exist without the star ratings, which allow for convenient sorting, filtering, and historical analysis over time (for instance, to track whether a restaurant is getting worse). This leads to what I'll term:

THE FIRST LAW OF INTERNET DATA
*In any computational context, explicitly structured data floats to the top.**

"Explicitly structured" data is any data that brings with it categories, quantification, and/or rankings. This data is self-contained, not requiring any greater context in order to be put to use. Data which exists in a structured and quantifiable context—be it the *DSM,* credit records, Dungeons & Dragons, financial transactions, Amazon product categories, or Facebook profiles—will become more useful and more important to algorithms—and to the people and companies using those algorithms—than unstructured data like text in human language, images, and video.

This law was obscured in the early days of the internet because there was so little explicitly quantified data there. Explicitly quantified *metadata* like the link graph, which Google exploited so lucratively, under-

* Like many things we call laws, these are soft laws—heuristics and guidelines—rather than inviolable iron laws.

scored that algorithms gravitate toward explicitly quantified data. In other words, *the early days of the internet were an aberration.* In retrospect, the early internet was the unrepresentative *beginning* of a process of explicit quantification that has since taken off with the advent of social media platforms like Facebook, Snapchat, Instagram, and Twitter, which are all part of the new norm.* This also includes Amazon, eBay, and other companies dealing in explicitly quantified data.

Web 2.0 was not about social per se. Rather, it was about the *classification* of the social, and more generally the classification of life. Google had vacuumed up all that could be vacuumed out of the unstructured data. The maturation of the web demanded more explicitly organized content that could more easily be analyzed by computers. And the best way to do this at scale was to employ users to create that data.

Explicitly quantified data requires that data be labeled and classified before it can be sorted and ordered. The project of archives like the Library of Congress isn't sorting the books per se; it's developing the overarching classification that determines what order the books should be in. No classification, no sorting. Even machine learning fares worse when "unsupervised"—that is, when it is not provided with a preexisting classificatory framework.

THE SECOND LAW OF INTERNET DATA

For any dataset, the classification is more important than what's being classified.†

The conclusions and impact of data analyses more often flow from the classifications under which the data has been gathered than from

* This is not to say that Facebook or other social engines will be the largest companies. Apple, Google, or other companies may continue to hold greater dominance in technology more generally, through their operating system platforms and other initiatives like self-driving cars and network infrastructure. In the online *social* sphere, however, the loose aggregative nature of Google has already given way to the current dominance of Facebook. Social networks regiment our *humanity* online.

† This law is an analogue to systems engineer and Go creator Rob Pike's fourth and fifth rules of software engineering, which emphasize *why* software engineering tends toward simplicity whenever possible: "Rule 4. Fancy algorithms are buggier than simple ones, and they're much harder to implement. Use simple algorithms as well as simple data structures. Rule 5. Data dominates. If you've chosen the right data structures and organized things well, the algorithms will almost always be self-evident. Data structures, not algorithms, are central to programming." Linus Torvalds expressed a similar sentiment: "Bad programmers worry about the code. Good programmers worry about data structures and their relationships."

the data itself. When Facebook groups people together in some category like "beer drinkers" or "fashion enthusiasts," there isn't some essential trait to what unifies the people in that group. Like Google's secret recipe, Facebook's classification has no actual secret to it. It is just an amalgam of all the individual factors that, when summed, happened to trip the category detector. Whatever it was that caused Facebook to decide I had an African-American "ethnic affinity" (was it my Sun Ra records?), it's not anything that would clearly cause a human to decide that I have such an affinity. What's important, instead, is that *such a category exists,* because it dictates how I will be treated in the future. The *name* of the category—whether "African American," "ethnic minority," "African descent," or "black"—is more important than the criteria for the category. Facebook's learned criteria for these categories would significantly overlap, yet the ultimate classification possesses a distinctly different meaning in each case. But the distinction between criteria is obscured. We never see the criteria, and very frequently this criteria is arbitrary or flat-out wrong. The choice of classification is more important than how the classification is performed.

Here, Facebook and other computational classifiers exacerbate the existing problems of provisional taxonomies. The categories of the *DSM* dictated more about how a patient population was seen than the underlying characteristics of each individual classified with these categories, because it was the category tallies that made it into the data syntheses. One's picture of the economy depends more on how unemployment is defined (whether it includes people who've stopped looking for a job, part-time workers, temporary workers, etc.) than it does on the raw experiences and opinions of citizens. And your opinion of your own health depends more on whether your weight, diet, and lifestyle are classified into "healthy" or "unhealthy" buckets than it does on the raw statistics themselves. Even the name of a category—"fat" vs. "overweight" vs. "obese"—carries with it associations that condition how the classification is interpreted.

Some classifications are far more successful and popular than others. The dominant rule of thumb is:

THE THIRD LAW OF INTERNET DATA

Simpler classifications will tend to defeat more elaborate classifications.

The simplicity of feedback mechanisms (Likes, star ratings, etc.) is intentional. Internet services *can* deal with complicated ontologies when they need to, but business and technical inertia privilege simpler ones. Facebook waited ten years to add reactions beyond "Like" and long resisted the calls for a "Dislike" button, forcing their users to Like death announcements and political scandals. Facebook preferred a simple bimodal interested/uninterested metric. When Facebook finally decided to appease their users, it added five sentiments to the original Like: Love, Haha, Wow, Sad, Angry. It is no coincidence that the two negative sentiments are at the end: "Sad" and "Angry" are more ambiguous than the others. If I express a positive reaction to something, I'm definitely interested in it. If I'm made sad or angry by something, I may still be interested in it, or perhaps I want to avoid it. Those reactions are less useful to Facebook.

Facebook's six reactions are similar to emoji in that they allow users to express emotion nonverbally, but they are more useful to Facebook *because* they comprise a simpler classification than the thousands of emoji. Viral content master Buzzfeed employs a similar, slightly hipper scheme for the reactions they permit users to post to their articles. Buzzfeed's scheme is tailor-made for market research: content can be surprising, adorable, shocking, funny, etc.

Bloomberg's Sarah Frier explained how Facebook formulated its new reactions:

> Facebook researchers started the project by compiling the most frequent responses people had to posts: "haha," "LOL," and "omg so funny" all went in the laughter category, for instance. . . . Then they boiled those categories into six common responses, which Facebook calls Reactions: angry, sad, wow, haha, yay, and love. . . . Yay was ultimately rejected because "it was not universally understood," says a Facebook spokesperson.

These primitive sentiments, ironically, enable more sophisticated analyses than a more complicated schema would allow—an important reason why simpler classifications tend to defeat more elaborate clas-

Buzzfeed's six possible reactions offered to article readers

sifications. Written comments on an article don't give Facebook a lot to go on; it's too difficult to derive sentiment from the ambiguities of written text unless the text is as simple as "lol" or "great." But a sixfold classification has multiple advantages. Facebook, Buzzfeed, and their kin seek universal and unambiguous sentiments. There is little to no variation in reaction choices across different countries, languages, and cultures.

The sentiments also make it easy to compare posts quantitatively: users themselves sort articles into "Funny," "Happy," "Sad," "Heartwarming," and "Infuriating." From looking at textual responses, it would be difficult to gauge that "Canada stalls on trade pact" and "Pop singer walks off stage" have anything in common, but if they both infuriate users enough to click the "Angry" icon, Facebook can detect a commonality. Those classifications permit Facebook to match users' sentiments with similarly classified articles, or try to cheer them up if they're sad or angry. If reactions to an article are split, Facebook can build subcategories like "Funny-Heartwarming" and "Heartwarming-Surprising." It can track which users react more with anger or laughter and so predict what kinds of content they'll tend to respond to in the future. Facebook can isolate particularly grumpy people and reduce their exposure to other users, so they don't drag down the Facebook population. And Facebook trains algorithms to make guesses about articles that don't yet have reactions. Most significantly, even though these particular six reactions are not a default and universal set, Facebook's choices will reinforce them as a default set, *making* them more

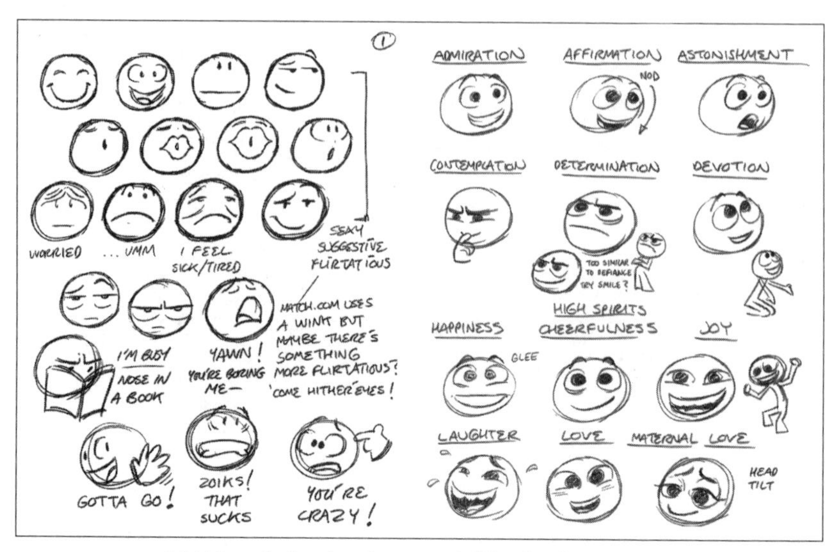

Matt Jones's sketches for potential Facebook Reactions

universal through a feedback loop.* The more we classify our reactions by that set of six, the more we'll be conditioned to gauge our emotions in those terms. The default six smooth out the variations observed when Facebook was conducting tests with a far larger set of emotions, all designed by Disney-Pixar's Matt Jones:

> The full list included admiration, affirmation, anger, anxiety, astonishment, awe, boredom, confusion, contemplation, contempt, contentment, coyness, curiosity, desire, determination, devotion, disagreement, disgust, embarrassment, enthusiasm, fear, gratitude, grief, guilt, happiness, high spirits, horror, ill temperament, indignant, interest, joy, laughter, love, maternal love, negation, obstinateness, pain, perplexity, pride, rage, relief, resignation, romantic love, sadness, shame, sneering, sulkiness, surprise, sympathy, terror, and weakness. . . . Clear patterns emerged in the data. Italians, South Africans, Russians and Brazilians had "Cultures of Love"—sending lots of amorous stickers. The U.S. and Canada were similar in most of their usage patterns—though

* *Wired* cheerfully observed that Facebook's Reactions "may not reflect the world in which we live, but they're a good deal closer to the one we want." This assumes that the world you want is one of standardized, superficial reactions to content devoid of nuance and sophistication.

Matt Jones's emotional sketches

the Canadians were vastly more "sympathetic," while the Americans were "sadder." And the use of "deadpan" stickers predominated across North Africa and the Middle East.

A simple classification won out. It is both easier to use and more universal—at the expense of cultural and personal variation.* And also, to hear researcher Dacher Keltner tell it to Radiolab's Andrew Zolli, at the expense of happiness:

> Countries that expressed the most "happiness" were not actually the happiest in real life. Instead, it was the countries that used the *widest array* of stickers that did better on various measures of societal health, well-being—even longevity. "It's not about being the happiest," Keltner told me, "it's about being the most emotionally diverse."

* On the other hand, Keltner and Facebook's research indicated that stereotypically happy Americans used the "Sad" Reaction more than nearly any other country save for those of the Middle East, Afghanistan, Pakistan, Mexico, Cambodia, Peru, Ireland, Azerbaijan, and a few others. Historically downbeat Russia used "Sad" Reactions significantly less than average. It seems hazardous to generalize from the use of the Reactions to people's actual emotional states. The types of responses may just not be as unified and universal as the Reactions make it seem.

The diversity comes not just from variety, but from ambiguity. One main appeal of emoji is that they aren't given to fixed definitions and leave room for interpretation, misinterpretation, and variant meanings across cultures and subcultures. It's hard to know what to make of this emoji, which goes by the name "upside-down face" (or U+1F643 officially):

Emojipedia describes the upside-down face as suggesting "silliness or goofiness. Sometimes used as an ambiguous emotion, such as joking or sarcasm." In other words, it can mean many things, depending on context—just like a *real* expression. If an upside-down smile is not ambiguous enough, there's the inscrutable "face without mouth" (U+1F636):

Here are some of the most popular emoji used on Twitter, as determined by the website Emojitracker:

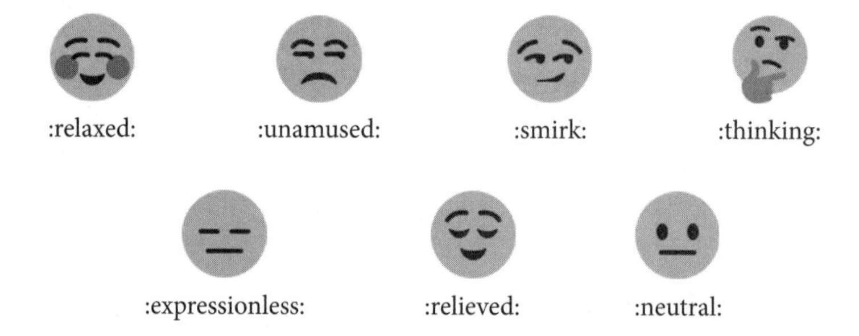

"Smirk" looks more sly than superior, while "neutral" and "expressionless" seem almost identical. "Thinking" (U+1F914) has become popularly used to signify sarcastic "throwing shade," to indicate vacuous disagreement. Providers like Facebook, Apple, Google, and mobile phone companies have their own designs that alter the nuance of the emoji, as with these variously constipated versions of "tired" (U+1F62B).

Different versions of the :tired: emoji

Over time, the meanings of emoji blur and diverge. The original Unicode characters for emoji were created to be compatible with the preexisting emoji that had originated in East Asia, which were then exported irregularly to the rest of the world. Such irregularities hampered data mining and sentiment analysis.

If the restricted, unambiguous set of six reactions has the effect of narrowing emotional diversity, social media and advertising companies view this tradeoff as the necessary cost of gathering better data on users. The restricted emotional language employed by Facebook is a language a computer can *understand* and *manipulate* at high scale. The simplified language of a core set of emotional reactions bridges the computational-human gap—more successfully than the overcomplicated ad hoc classifications of the *DSM* did. Instead, these reaction sets are reminiscent of the simpler folk taxonomies of Myers-Briggs, OCEAN, and HEXACO, which also break down complex phenomena onto a handful of axes. Facebook's Reactions even approximately map to the Big Five:

Like: Agreeableness
Love: Extroversion
Wow: Openness
Sad: Neuroticism
Angry: Conscientiousness

The odd one out is Haha, because as always, laughter eludes easy classification despite being the most universal and nonnegotiable of expressions. Yet for the remaining five, there is an inevitable flattening of cultural differences. We saw how much trouble personality classification efforts had while trying to generalize their models across different cultures, and despite Facebook's empirical research to generalize their six, it's unlikely that they're capturing the *same* sentiments across cultures—rather, they found sentiments that were *recognizable* by multiple cultures. If the data miners and user profilers get their way, soon enough we will all be Loving, Wowing, Sadding, and Angrying in lockstep.

The language of Reactions is a primitive vocabulary of emotion, vastly simpler than our human languages. It is far better suited to computers and computational analysis. When I introduced graphical emoticons into the Messenger client in 1999, I didn't foresee any of this. Around 2015, I began noticing a change on my Facebook wall. There was less discussion happening. People I knew were far more content to respond to posts with monosyllables like "yeah" or "ugh," or else with simple emoji or Facebook's six reactions. I caught myself contributing dumbly like this, to my dismay. I went back and checked posts from 2009 and 2010. I had written in complete sentences, arguments, with multiple paragraphs. The shift was obvious and drastic. Diversity, nuance, and ambiguity had declined. If passions were fervent and I disagreed with the chorus of "yeahs" or "ughs," the crowd was far more likely to pounce on me; the same went for any other dissenters. What had happened? *These were my friends.* But they no longer seemed like the same people. We had been standardized. We were all speaking a different language now. It was the language of Facebook—of computers.

EPILOGUE: THE REDUCTION OF LANGUAGE, THE FLATTENING OF LIFE

> I bought translations of all kinds of my own existence.
>
> —TOM PHILLIPS, *A Humument*

COMPOSER Alvin Lucier's masterpiece of minimalism, "I Am Sitting in a Room," begins with the following text, spoken into a tape recorder:

> I am sitting in a room different from the one you are in now. I am recording the sound of my speaking voice and I am going to play it back into the room again and again until the resonant frequencies of the room reinforce themselves so that any semblance of my speech, with perhaps the exception of r-r-r-rhythm, is destroyed. What you will hear, then, are the natural resonant frequencies of the room articulated by speech. I regard this activity nnnnnot so much as a demonstration of a physical fact, but more as a way to s-s-smooth out any irregularities my speech might have.

Quickly, Lucier's words become incomprehensible. He uses the language of combat—"reinforce," "destroyed." What occurs is a battle in which the feedback process overcomes the spoken words. The room slowly triumphs, rendering soft, metallic feedback in place of speech. What survives are the aspects of Lucier's speech that were most compatible with the room. Lucier's words, his voice, and his stutter are lost to those aspects that the room most strongly reflects. Instead of words, there are frequencies, just as computers take our words and reduce them to coded numbers, grouped en masse.

Attuned to the new digital spaces in which we exist, we are slowly building a hybridized language that bridges human and machine. This language empowers us to better analyze the resulting data, but it also eliminates our irregularities—that which is lost in the translation to its computational representation. Machines play back to us smoothed and regularized versions of our lives and interactions, precisely encoded and quantified. These are inexact replicas of more complicated phenomena, and yet for the benefits of technology, we accept computers' version of us as the real thing. Irregularities are treated as pathological exceptions, or else they're ignored altogether.

The playwright Richard Foreman declared that humans are becoming "pancake people . . . spread wide and thin as we connect with that vast network of information accessed by the mere touch of a button." I think that this flattening, or smoothing as Lucier would have it, is indeed taking place, but it is not caused intrinsically by the glut of information surrounding us. Rather, it is caused by a feedback process: we see approximated digital images of ourselves and we take them to be who we truly are. We begin to think we are the room. Our information today is very different from the information we had fifty or even twenty years ago, because it is more compatible with computers. It is the language of labels and classifications, a lingua franca for computers and humans alike.

I am, in general, cautiously positive toward technology. I don't have faith that technology, broadly speaking, is guaranteed to produce a greater good. Rather, I believe that we signed a Faustian contract at the dawn of the modern age. I believe we can address the wide-scale catastrophes threatened by global warming, unsustainable ecology, and the possible failure of our increasingly complex infrastructure only through the development of further, more advanced technologies. We may blow ourselves up before we save ourselves, but I don't consider halting or reversing technological progress to be a plausible option. Science and technology are terrifying yet amazing, but they also produce a dangerous amnesia as change accelerates. I've felt it myself.

I would not trade my life for the life of the most well-off person in ancient Greece, Renaissance England, or Enlightenment France—three periods with which I'm familiar enough to have affection for. But in even the most barbarous and oppressive of historical moments, there remain aspects of past human experience that are edifying, beautiful, and even occasionally uplifting. This too is knowledge, and it is the sort of knowledge that is most likely to be lost in the ever-growing move toward capturing human experience in computation. This knowledge can be only approximately communicated through human language, so difficult for computers to grasp.

In fearing this loss, I am a conservative in the way that the Marxist philosopher Gerald Cohen described himself:

> The conservative propensity is to conserve, to not destroy, and, therefore, to not replace, even (within limits) by something more valuable. A conservative can believe that what rises from the ashes is the greatest building ever and that it was right to build it, yet still feel distraught that the old building was destroyed.

I was granted an opportunity to be an agent of the technological change that we are currently undertaking, and I greatly enjoyed it. I took it as a refuge from the irrational complications and irregularities that computers ignore. Those irregularities—whether in spoken language or in literature or in my own children—nonetheless hold deep and complex meaning for me. I do not wish for them to be lost in a steamroller of standardization. Computers will not grasp these depths of meaning for decades, perhaps centuries. I desire for those edifices of art and feeling to be preserved in the intervening years. I believe in their benefit to humanity and in their innate positive value. My faith is that we need our unqualified nuances and irregularities to elude code and ontology. We need them to be recognized and given meaning by the world, however computationalized it becomes.

ACKNOWLEDGMENTS

I am grateful for the insight and support that so many have provided me over my life. I am fortunate to have known all those below.

My teachers: Cathy Fiore, John Clyman, Jim Holland, James Parkman, Doug Biedenweg, Tony Hodgin, Tom Generous, David Doster, William Flesch, John Crowley, Heinrich von Staden, David Pears, Derk Pereboom, Paolo Mancosu, Stanley Eisenstat, Zhong Shao, Edmund L. Epstein, Galen Strawson, David Rosenthal, David Carr. I learned more from you than you know.

My friends and colleagues who enriched my time at Google and Microsoft: Chris Pirich, Jonathan Forbes, William Lai, Bama Rao, Kara Lewis, Chris Mitchell, Yikang Xu, David Anson, Lan Tang, Richard Chung, Niniane Wang, Larry Greenfield, Arup Mukherjee, Joanna Kulik, Daniel Dalitz, Debby Wallach, Peter Weinberger, Bill Coughran.

Fellow writers whose existence encourages me: Jessie Ferguson, Paul Kerschen, Ray Davis, Juliet Clark, Kathryn Hume, Annie Kim, Lisa Samuels, Peli Grietzer, Susan Bernofsky, Stephen Dixon, and László Krasznahorkai.

For intense debate and engagement with my inchoate ideas: James Grimmelmann, Josh Harrison, Adam Elkus, Brett Fujioka, Josephine Wolff, Joel Hernandez, John Emerson, Natasha Singer, Frank Pasquale, Cosma Shalizi, Henry Farrell, Meredith L. Patterson.

For necessary support in the wild world of publishing: Alexander Provan, Sam Frank, Dan Visel, Keith Gessen, Marco Roth, Carla Blumenkranz, Dayna Tortorici, Jessica Winter, Laura Helmuth, Jill Schoolman, Harry Siegel.

Friends old and new: Laura Skorina, Harrison Hung, Gabriella

Gruder-Poni, Jaya Kasibhatla, Seo-Young Chu, Cary Franklin, Cori Gabbard, Kaveri Nair, Lauren Gabriele, Chris Hardgrove, Anna Medvedovsky, Noémie Elhadad, Mercedes Armillas, Jenny Chisnell, Anastasia Senenko, Mary McMullen, Eleanor Sarasohn, Jannon Stein, Hari Khalsa, Kelly Molloy, Florence Liu, Blakely Phillips, Erica Weitzman, Simona Sivkoff, Tal Corem, Katie Sigelman, Scott Brown, Maria Schurr, Leigh Fullmer, Sophie Rollins, Lauren Moos, Sara Plourde, Stephen Lavelle, Sonya Mann, Heather Day-Richter, Cynthia Campos, Eireene Nealand, Juliet O'Keefe, Yi Shi.

Cognitive expertise: Stan Smith, Angelica Kaner, Trip Quillman, Ken Corbett, Ron Winchel, Virginia Goldner.

For musical sustenance in a time of need: Peggy Sartoris-Belacqua, Henry Hughes, and Jon Abbey.

For encouragement above and beyond: Jordan Ellenberg, Scott Aaronson.

For the crucial support in this book's existence: Sarah Burnes, Dan Frank, Maria Goldverg, Nora Reichard, Daniel Seidel.

For their generosity in allowing their work to be used: Richard McGuire, Brecht Evens, Tim Denee, Matt Jones, Craig VanGrasstek.

And of course to my family, who made me who I am. Nina, this book would not exist without your unceasing support and insight.

The writing of this book was supported by a fellowship from New America. They have not exerted any undue influence over its contents.

NOTES

1 LOGO AND LOVE

9 I found particular pleasure: Papert, 1980, vi–vii.
15 We walk through the world: Reichenbach, 1938.
16 Roger Cramton attributes it: Cramton, 1986.
20 "With the accessibility of music": Cope, 2003.
24 COUT gequ $FDED: Shepherd.
27 He was a one-man: Auerbach, "A Delville of a Tolkar: Martin Gardner's 'Undiluted Hocus-Pocus,'" 2013.
27 Georges Perec's mighty *Life*: Mathews & Brotchie, 2005.
28 Perec used (and abused) Graeco-Latin squares: Bellos, 1993, 596–608.
29 Perec, unable to construct squares: Ibid., 420.
29 "perhaps the greatest 20th century novel": Knuth, n.d.
30 The modern sense of the term "heuristic": Pólya, 1957.
31 "The one sort are above all": Poincaré, 1907, 15–21.
31 Seventy years later: Kac, 1985, xxv.
31 "Without bias": Gigerenzer, Hertwig, & Pachur, 2011.
34 One version: Gilbert, 1955.
42 "Stoner looked across": Williams J., 1965.

2 CHAT WARS

48 "In Cyberspace": Hansell, 1999.
48 Despite Microsoft's notorious: Auerbach, "Tales of an Ex-Microsoft Manager," 2013.
49 Following the far less popular Lisa: Pang.
50 More significantly: Rosenoer, 1997, 49–53.
50 "Almost all the similarities": *Apple Computer, Inc. v. Microsoft Corp.,* 1994.
53 "I don't want to be remembered": Bank, 2001.
53 "Anyone who has had the misfortune": Truesdell, 1984, 629.
54 I built this feature: Auerbach, "I Built That 'So-and-So Is Typing' Feature in Chat," 2014.

59 "Mr. Smith": Chappell, 2008.
61 When Windows architect: Mazarakis & Shontell, 2017.

3 BINARIES

63 "Our ordinary habits": Naur, 1990.
63 "starry dance": Milton, *Paradise Lost,* III.579–81.
67 "Without any information": Smullyan, 1978, 65.
70 "The letters on a computer screen": Associated Press, 1999.
71 "Those who say mathematics": Ramsey, 1931.
75 "To say of what is": Aristotle, 1908, 1011b25.
75 "Mathematicians are therefore mystified": Rota, 1997/2008, 93.
75 "Truth, in Plato's system": Friedlander, 1969, 227.
75 "Truth *happens*": James, 1907.
75 "We feel that there is an inner kinship": Carnap, 1967/2003, xviii.
77 "If we translate 'scientific outlook' ": Musil, 1995.
78 "We are unable clearly": Wittgenstein, 1960, 25.
79 We invent and create: Auerbach, "The Limits of Language," 2015.

INTERLUDE: FOREIGN TONGUES

82 "Deeply lost": Kafka, 1971.
82 "Speech then is not": Kleist, 1997/2004, 408–9.
84 "While LiveJournal": Patterson, 2013.
84 "I find in writers": Diderot, 2006, 49.

4 NAMING OF PARTS

94 That is, one's gender: World Health Organization, 2015.
94 If, as the World Health Organization states: Ibid.
99 "In some sciences": Lichtenberg, 2012.
102 The MBTI publisher: Winterhalter, 2014.
105 Read enough of this bombast: "Well-Known Rationals," n.d.

5 SELF-APPROXIMATIONS

113 "Putting facts": Sapolsky, 2017.
113 the "Big Five": Spielman, 2016, 391.
114 the Big Five model: McCrae & Costa, 2003.
114 "Most constructs": Nettle, 2007, 10, 39.
115 "Ashton's model basically divides": McCrae, *Cambridge Handbook of Personality Psychology,* 2009.
117 "It would be indeed unusual": Lem, 2013.
129 "The business model": Frances, 2013.

129 "Take the case of Dr. Joseph": Angell, 2009.
130 "We saw *DSM-IV*": Frances, 2013.
131 "Boys who were born": Morrow et al., 2012.
131 As family therapist: Lipuscek, 2016.
131 Are mania, depression: Shorter, 2015, 167–70.
131 "The right goal": Frances, 2013.
132 "As documented": Schatzberg, Scully Jr., Kupfer, & Regier, 2009.
132 "Our immediate task": Regier, Narrow, Kuhl, & Kupfer, 2009.
132 "The challenge for *DSM-V*": Carpenter, 2009.
133 "a new set of advantages": Helzer, Kraemer, & Krueger, 2006.
133 "Perhaps the greatest": Ibid.
133 "Consistency in the collection": Ibid.
134 "a direct, dimensional reflection": Ibid.
134 Psychotherapist Gary Greenberg's account: Greenberg, 2013.
135 "While *DSM* has been described": Insel, 2013.
135 "There's no reality": Greenberg, 2013.

6 GAMES COMPUTERS PLAY

136 "What is it that you see": Pagels, 1988.
141 The *Minneapa* fanzine's: Peterson, *Playing at the World,* 2012.
143 "will insist upon everything": Ibid.
143 As the hobbyist: Rossiter, 2012, 61.
145 dipsomania: Gygax, 1979.
146 "Schizophrenia": Ibid.
147 "Essential D&D": Kruger, 2016.
147 "to discover a persona": Peterson, *Playing at the World,* 2012.
147 "When an individual": Goffman, *The Presentation of Self in Everyday Life,* 1956.
149 "A dozen tries later": Aaron, 1985.
150 "Suppose that a crisis": Crawford, 1986/2014.
153 *Dwarf Fortress*'s more realistic: StarkRavingMad, 2007.
154 A procedurally generated: Sankis, 2007.
154 "To begin with, all": StarkRavingMad, 2007.
156 "Once, a town's executioner": Dale, 2014.
156 "Do we do law": Fenlon, 2016.
157 "I added taverns": Ibid.
158 "There are so many interlocking systems": Ibid.
159 "Every program": Perlis, 1982.
159 "use the rules only": Kruger, 2016.

INTERLUDE: ADVENTURES WITH TEXT

161 I remember *Balance of Power:* Auerbach, "The Hardest Computer Game of All Time," 2014.
166 You begin to approach: Moriarty, 1986.
171 "Without puzzles": Nelson, 2001, 382.

171 examine machine: Ibid.
173 Some authors: Maher, 2014.

7 BIG DATA

180 "Simple models": Halevy, Norvig, & Pereira, 2009.
183 In 2016, "best mesothelioma lawyer": Lake, 2016.
188 Google had hoped: Grimmelmann, "Hail and Farewell to the Google Books Case," 2016, and Grimmelmann, "Eight Years Later, the Google Books Fight Lumbers On," 2013.
188 This cutoff coincides: Carlisle, 2014.
189 "All that it is given": Borges, 1999.
189 "[The Company's] silent functioning": Ibid.
192 The guidelines, it turned out: "Trending Review Guidelines," 2016.
194 Mathematician Godfrey: Hardy, 1940/2005, 1.
195 "But is not the position": Ibid., 41.
195 "Judged by all": Ibid., 49.
197 "I had been calling them": Gell-Mann, "Naming Quarks," 1997.
197 "Joyce built his house": Glasheen, 1977, xi–xiii.

8 PROGRAMMING MY CHILD

201 "When adults at 4 months": Reddy, 2008.
206 "Adjustment to objective reality": Vygotsky, *Thought and Language,* 1986, 37.
207 "For example, preschoolers": Gelman, S., 2003, 122.
208 "Essentialism is not": Ibid., 295.
209 "[A child] begins to find": Peirce, 1868.
210 Physicist Juan Roederer: Roederer, 2005.
212 The rhetoric around: Shapiro & Forrest, 2016.
213 "It may very well be": Shanon, 2003.
213 "MDMA administration": Danforth et al., 2015.
214 "Why is the mind": McCulloch, 1951/1988.
214 "We don't need something more": Gell-Mann, *Nature Conformable to Herself,* 2010, 381.
219 As Susan Gelman found: Gelman, S., 2003, 31–33.
220 "thought becomes verbal": Vygotsky, *Thought and Language,* 1986, 83.
220 "The meaningful word": Vygotsky, *The Collected Works of L. S. Vygotsky: Problems of General Psychology,* 1987, 285.

9 BIG HUMAN

223 "Technology's primary effect": Toyama, 2015.
224 Beginning in 2000: Woods, 2016.
226 At the time of the September 11 attacks: Johnston & Lewis, 2002.

226 "Alexander wants": Harris, 2013.
227 "Two engineers": Gellman & Soltani, 2013.
227 A 2010 *Washington Post:* Ibid., and Kaplan, 2016, 152–57.
227 Later documents leaked: Auerbach, "MUSCULAR 'Roid Rage!," 2013.
227 A 2010 UK report: "The Digint Programme," 2016.
227 "Alexander asked": Bamford, 2016.
228 "In a top-secret memo": Ibid.
229 Consumer profiling: Singer, "Mapping, and Sharing, the Consumer Genome," 2012, and Turow, 2012.
229 1. Location: Dewey, 2016; Escobar, 2017.
237 The link took me: Alciné, 2015.
238 One user posted a photo: Griffin, 2015.
238 "Certain offensive search terms": Fitzpatrick, 2015.
238 In practice, this equated: Sullivan, 2015.
239 "the angle θ": Wu & Zhang, 2016.
239 A substantial portion: Open Science Collaboration, 2015.
239 "Although criminals": Wu & Zhang, 2016.
240 "Relatively little study": Ibid.
241 "By definition, of course": Goffman, *Stigma: Notes on the Management of Spoiled Identity,* 1963.
245 The field of "sentiment analysis": Grohol, 2014.
246 Facebook backed off tracking efforts: Auerbach, "You Are What You Click," 2013.
248 "Rule 4. Fancy algorithms": Pike, 1989.
248 "Bad programmers worry": Torvalds, 2006.
250 "Facebook researchers started": Frier, 2016.
252 *Wired* cheerfully observed: Gonzalez, 2015.
252 "The full list included": Zolli, 2015.
253 "It's not about being": Ibid.

EPILOGUE: THE REDUCTION OF LANGUAGE, THE FLATTENING OF LIFE

258 "pancake people": Foreman, 2005.
259 The conservative propensity: Cohen, 2011.

FURTHER READING

On the history of computing: Subrata Dasgupta's *It Began with Babbage,* Margaret Boden's *Mind as Machine: A History of Cognitive Science,* Mark Priestley's *A Science of Operations: Machines, Logic and the Invention of Programming,* Edgar Daylight's *The Dawn of Software Engineering: From Turing to Dijkstra.*

On computer science: John MacCormick's *9 Algorithms That Changed the Future,* Charles Petzold's *Code,* Peter J. Denning and Craig H. Martell's *Great Principles of Computing,* David Harel's *Algorithmics,* Scott Aaronson's *Quantum Computing Since Democritus.*

Reflections on science: Henri Poincaré's *The Value of Science,* Jacob Bronowski's *The Origins of Knowledge and Imagination,* Gian-Carlo Rota's *Indiscrete Thoughts,* Heinz Pagels's *The Dreams of Reason,* Jeremy Gray's *Plato's Ghost: The Modernist Transformation of Mathematics.*

On philosophy and language: Raymond Smullyan's *This Book Needs No Title,* Denis Diderot's *D'Alembert's Dream,* Albert Atkin's *Peirce,* Ludwig Wittgenstein's *Philosophical Investigations,* David Stern's *Wittgenstein on Mind and Language,* Laura Riding's *Anarchism Is Not Enough,* James O'Shea's *Wilfrid Sellars: Naturalism with a Normative Turn.*

On James Joyce: Frank Budgen's *James Joyce and the Making of Ulysses,* Roland McHugh's *The Finnegans Wake Experience,* Philip Kitcher's *Joyce's Kaleidoscope: An Invitation to Finnegans Wake.*

On role-playing games: Jon Peterson's *Playing at the World.*

WORKS CITED

Aaron, D. "Playing with Apocalypse." *New York Times,* December 29, 1985. http://www.nytimes.com/1985/12/29/magazine/playing-with-apocalypse.html?pagewanted=all.

Abelson, H., & Sussman, G. J. *The Structure and Interpretation of Computer Programs* (2nd ed.). Cambridge: MIT Press, 1996.

Alciné, J. Twitter. June 28, 2015. Retrieved from https://twitter.com/jackyalcine/status/615329515909156865.

Angell, M. "Drug Companies & Doctors: A Story of Corruption." *New York Review of Books,* January 15, 2009. Retrieved from http://www.nybooks.com/articles/2009/01/15/drug-companies-doctorsa-story-of-corruption.

Apple Computer, Inc. v. Microsoft Corp., 35 F.3d 1435. Court of Appeals, 9th Circuit, September 19, 1994. https://scholar.google.com/scholar_case?case=17794375458513139314.

Aristotle. *The Works of Aristotle: Metaphysics* (vol. 8). Translated by W. D. Ross. Oxford: Clarendon Press, 1908.

Associated Press. "Rabbi OKs Deleting 'God' on Computers." *Los Angeles Times,* January 2, 1999.

Auerbach, D. "A Delville of a Tolkar: Martin Gardner's 'Undiluted Hocus-Pocus.'" *Los Angeles Review of Books,* November 4, 2013. Retrieved from https://lareviewofbooks.org/article/a-delville-of-a-tolkar-martin-gardners-undiluted-hocus-pocus/.

———. "MUSCULAR 'Roid Rage!" *Slate,* October 31, 2013. Retrieved from http://www.slate.com/articles/technology/bitwise/2013/10/nsa_muscular_program_spying_on_google_and_yahoo.html.

———. "Tales of an Ex-Microsoft Manager." *Slate,* August 26, 2013. Retrieved from http://www.slate.com/articles/business/moneybox/2013/08/microsoft_ceo_steve_ballmer_retires_a_firsthand_account_of_the_company_s.html.

———. "You Are What You Click." *The Nation,* March 4, 2013. http://www.thenation.com/article/you-are-what-you-click-microtargeting/.

———. "I Built That 'So-and-So Is Typing' Feature in Chat." *Slate,* February 14, 2014. Retrieved from http://www.slate.com/articles/technology/bitwise/2014/02/typing_indicator_in_chat_i_built_it_and_i_m_not_sorry.html.

———. "The Hardest Computer Game of All Time." *Slate,* January 24, 2014. Retrieved from http://www.slate.com/articles/technology/bitwise/2014/01/robot_odyssey_the_hardest_computer_game_of_all_time.html.

———. "The Limits of Language." *Slate,* September 1, 2015. Retrieved from http://www.slate.com/articles/life/classes/2015/09/take_a_wittgenstein_class_he_explains_the_problems_of_translating_language.html.

Bamford, J. "Every Move You Make." *Foreign Policy,* September 7, 2016. Retrieved from https://foreignpolicy.com/2016/09/07/every-move-you-make-obama-nsa-security-surveillance-spying-intelligence-snowden/.

Bank, D. *Breaking Windows: How Bill Gates Fumbled the Future of Microsoft.* New York: Free Press, 2001.

Bellos, D. *Georges Perec: A Life in Words.* Boston: David R. Godine, 1993.

Bickhard, M. H. *Cognition, Convention, and Communication.* New York: Praeger, 1980.

Borges, J. L. *Collected Fictions.* Translated by A. Hurley. New York: Penguin Books, 1999.

Carlisle, S. "Mickey's Headed to the Public Domain! But Will He Go Quietly?" Nova Southeastern University, October 17, 2014. Retrieved from http://copyright.nova.edu/mickey-public-domain/.

Carnap, R. *The Logical Structure of the World; and, Pseudoproblems in Philosophy.* Translated by R. A. George. Chicago: Open Court, 1967/2003.

Carpenter, W. T. "Criticism vs Fact: A Response to a Warning Sign on the Road to *DSM-V* by Allen Frances, MD." *Psychiatric Times,* July 27, 2009. Retrieved from http://www.psychiatrictimes.com/articles/criticism-vs-fact-response-warning-sign-road-dsm-v-allen-frances-md.

Chappell, G. "The Phil Bucking Correspondence," January 26, 2008. Retrieved from http://www.geoffchappell.com/notes/security/aim/bucking.htm.

Cohen, G. A. "Rescuing Conservatism: A Defense of Existing Value." In *Reasons and Recognitions: Essays on the Philosophy of T. M. Scanlon,* edited by R. J. Wallace, R. Kumar, & S. Freeman, 203–30. Oxford: Oxford University Press, 2011.

Cope, J. "Mirrors." Head Heritage, July 2003. Retrieved from https://www.headheritage.co.uk/unsung/albumofthemonth/mirrors.

Cramton, R. C. "Demystifying Legal Scholarship." *Georgetown Law Journal* 75 (1987): 1–2. http://scholarship.law.cornell.edu/facpub/1006.

Crawford, C. *Balance of Power: The Book,* 1986/2014. Retrieved from http://www.erasmatazz.com/MyResources/Balance-of-Power-the-Book.pdf.

Dale, B. "The Adams Brothers Behind 'Dwarf Fortress' Game Are Weirdly Brilliant." Technical.ly, May 21, 2014. Retrieved from http://technical.ly/brooklyn/2014/05/21/dwarf-fortress-adams/.

Danforth, A., et al. "MDMA-Assisted Therapy: A New Treatment Model for Social Anxiety in Autistic Adults." *Progress in Neuro-Psychopharmacology & Biological Psychiatry* (2015): 237–49. https://doi.org/10.1016/j.pnpbp.2015.03.011.

Daylight, E. G. *The Dawn of Software Engineering: From Turing to Dijkstra.* Heverlee, Belgium: Lonely Scholar, 2012.

Dewey, C. "98 Personal Data Points That Facebook Uses to Target Ads to You." *Washington Post,* August 19, 2016. Retrieved from https://www.washingtonpost.com/news/the-intersect/wp/2016/08/19/98-personal-data-points-that-facebook-uses-to-target-ads-to-you/.

Diderot, D. *Rameau's Nephew and First Satire.* Translated by M. Mauldon. Oxford: Oxford University Press, 2006.

Escobar, M. "Facebook Advertising Detailed Targeting Options." July 20, 2017. Retrieved from http://www.edigitalagency.com.au/facebook/facebook-advertising-targeting-options/.

Fenlon, W. "Dwarf Fortress' Creator on How He's 42% Towards Simulating Existence." *PC Gamer,* March 31, 2016. Retrieved from http://www.pcgamer.com/dwarf-fortress-creator-on-how-hes-42-towards-simulating-existence/2/.

Fitzpatrick, J. "Sorry for Our Google Maps Search Mess Up." Google, May 21, 2015. Retrieved from https://maps.googleblog.com/2015/05/sorry-for-our-google-maps-search-mess-up.html.

Foreman, R. "The Pancake People, Or, 'The Gods Are Pounding My Head.'" Edge, March 8, 2005. Retrieved from https://www.edge.org/3rd_culture/foreman05/foreman05_index.html.

Frances, A. *Saving Normal.* New York: William Morrow, 2013.

Friedlander, P. *Plato: An Introduction* (2nd ed.). Translated by H. Meyerhoff. Princeton, NJ: Princeton University Press, 1969.

Frier, S. "Inside Facebook's Decision to Blow Up the Like Button." *Bloomberg,* January 27, 2016. Retrieved from http://www.bloomberg.com/features/2016-facebook-reactions-chris-cox/.

Gellman, B., & Soltani, A. "NSA Infiltrates Links to Yahoo, Google Data Centers Worldwide, Snowden Documents Say." *Washington Post,* October 30, 2016. Retrieved from https://www.washingtonpost.com/world/national-security/nsa-infiltrates-links-to-yahoo-google-data-centers-worldwide-snowden-documents-say/2013/10/30/e51d661e-4166-11e3-8b74-d89d714ca4dd_story.html.

Gell-Mann, M. "Naming Quarks." Interview by G. West. Web of Stories, October 1997. Retrieved from https://www.webofstories.com/play/murray.gell-mann/116.

———. "Nature Conformable to Herself." In *Selected Papers,* edited by H. Fritzsch, 378–81. Hackensack, NJ: World Scientific, 2010.

Gelman, A., Fagan, J., & Kiss, A. "An Analysis of the New York City Police Department's 'Stop-and-Frisk' Policy in the Context of Claims of Racial Bias." *Journal of the American Statistical Association* 102, no. 479 (September 2007): 813–23. https://doi.org/10.1198/016214506000001040.

Gelman, S. *The Essential Child: Origins of Essentialism in Everyday Thought.* Oxford: Oxford University Press, 2003.

Gigerenzer, G., Hertwig, R., & Pachur, A. T. *Heuristics: The Foundations of Adaptive Behavior.* Oxford: Oxford University Press, 2011.

Gilbert, S. *James Joyce's Ulysses: A Study* (2nd ed.). New York: Vintage, 1955.

Glasheen, A. *A Third Census of Finnegans Wake.* Berkeley: University of California Press, 1977.

Goffman, E. *The Presentation of Self in Everyday Life.* Edinburgh: University of Edinburgh Social Sciences Research Center, 1956.

———. *Stigma: Notes on the Management of Spoiled Identity.* New York: Penguin, 1963.

Gonzalez, R. "The Surprisingly Complex Design of Facebook's New Emoji." *Wired,* October 12, 2015. Retrieved from https://www.wired.com/2015/10/facebook-reactions-design/.

Greenberg, G. *The Book of Woe: The DSM and the Unmaking of Psychiatry.* New York: Blue Rider Press, 2013.

Griffin, A. "Flickr's Auto-Tagging Feature Goes Awry, Accidentally Tags Black People as Apes." *The Independent,* May 20, 2015. Retrieved from http://www.independent.co.uk/life-style/gadgets-and-tech/news/flickr-s-auto-tagging-feature-goes-awry-accidentally-tags-black-people-as-apes-10264144.html.

Grimmelmann, J. "Eight Years Later, the Google Books Fight Lumbers On." *Publishers Weekly,* September 5, 2013. Retrieved from https://www.publishersweekly.com/pw/by-topic/digital/content-and-e-books/article/58953-eight-years-later-the-google-books-fight-lumbers-on.html.

———. "Hail and Farewell to The Google Books Case." *Publishers Weekly,* May 11, 2016. Retrieved from https://www.publishersweekly.com/pw/by-topic/digital/copyright/article/70326-hail-and-farewell-to-the-google-books-case.html.

Grohol, J. M. "Emotional Contagion on Facebook? More Like Bad Research Methods." Psych Central, June 23, 2014. Retrieved from http://psychcentral.com/blog/archives/2014/06/23/emotional-contagion-on-facebook-more-like-bad-research-methods/.

Gygax, G. *Dungeon Master's Guide.* Lake Geneva, WI: TSR, 1979.

Halevy, A., Norvig, P., & Pereira, F. "The Unreasonable Effectiveness of Data." *IEEE Intelligent Systems* 24, no. 2 (2009): 8–12. https://doi.org/10.1109/MIS.2009.36.

Hansell, S. "In Cyberspace, Rivals Skirmish Over Messaging." *New York Times,* July 24, 1999. http://www.nytimes.com/1999/07/24/business/in-cyberspace-rivals-skirmish-over-messaging.html.

Hardy, G. H. *A Mathematician's Apology.* Edmonton: University of Alberta Mathematical Sciences Society, 1940/2005. Retrieved from http://www.math.ualberta.ca/mss/misc/A%20Mathematician's%20Apology.pdf.

Harris, S. "The Cowboy of the NSA." *Foreign Policy,* September 9, 2013. Retrieved from http://foreignpolicy.com/2013/09/09/the-cowboy-of-the-nsa/.

Harvey, P. *An Introduction to Buddhism: Teachings, History and Practices* (2nd ed.). Cambridge: Cambridge University Press, 2013.

Helzer, J. E., Kraemer, H. C., & Krueger, R. F. "The Feasibility and Need for Dimensional Psychiatric Diagnoses." *Psychological Medicine* 36, no. 12 (2006): 1671–80. https://doi.org/10.1017/S003329170600821X.

Insel, T. "Transforming Diagnosis." National Institute of Mental Health, April 29, 2013. Retrieved from https://www.nimh.nih.gov/about/directors/thomas-insel/blog/2013/transforming-diagnosis.shtml.

James, W. "Pragmatism's Conception of Truth." In *Pragmatism: A New Name for Some Old Ways of Thinking,* 76–91. New York: Longman Green, 1907.

Johnston, D., & Lewis, N. A. "Whistle-Blower Recounts Faults Inside the F.B.I." *New York Times,* June 7, 2002. http://www.nytimes.com/2002/06/07/us/traces-terror-congressional-hearings-whistle-blower-recounts-faults-inside-fbi.html.

"Judging or Perceiving." The Myers & Briggs Foundation, n.d. Retrieved from http://www

.myersbriggs.org/my-mbti-personality-type/mbti-basics/judging-or-perceiving.htm.

Kac, M. *Enigmas of Chance: An Autobiography.* Berkeley: University of California Press, 1985.

Kafka, F. "At Night." Translated by T. A. Stern. In *The Complete Stories,* edited by N. N. Glatzer, 436. New York: Schocken Books, 1971.

Kaplan, F. *Dark Territory: The Secret History of Cyber War.* New York: Simon & Schuster, 2016.

Kleist, H. v. "On the Gradual Production of Thoughts Whilst Speaking." Translated by P. Constantine. In *Selected Writings,* edited by P. Constantine, 405–9. Indianapolis, IN: Hackett, 1997/2004.

Knuth, D. "Retirement." n.d. Retrieved from http://www-cs-faculty.stanford.edu/~uno/retd.html.

Kruger, B. "Bob Kruger on the Essentials of D&D." The BusyBody, January 18, 2016. Retrieved from https://rossonl.wordpress.com/2016/01/18/guest-blogger-bob-kruger-on-the-essentials-of-dd/.

Lake, C. "The Most Expensive 100 Google Adwords Keywords in the US." Search Engine Watch, May 31, 2016. Retrieved from https://searchenginewatch.com/2016/05/31/the-most-expensive-100-google-adwords-keywords-in-the-us/.

Lem, S. *Summa Technologiae.* Translated by J. Zylinska. Minneapolis: University of Minnesota Press, 2013.

Lichtenberg, G. C. *Philosophical Writings.* Translated by S. Tester. Albany: State University of New York Press, 2012.

Lipuscek, J. "Youngest Children in Class at Increased Risk of ADHD Diagnosis & Medication." Joan Lipuscek, May 8, 2016. Retrieved from https://www.joanlipuscek.com/child-therapy-news/youngest-children-in-class-at-greater-risk-for-adhd.

Maher, J. "Mindwheel (or, The Poet and the Hackers)." The Digital Antiquarian, March 10, 2014. Retrieved from http://www.filfre.net/2014/03/mindwhell-or-the-poet-and-the-hackers/.

Mathews, H., & Brotchie, A, eds. *Oulipo Compendium* (2nd ed.). London: Atlas Press, 2005.

Mazarakis, A., & Shontell, A. "Steve Ballmer Says He Never Actually Threw a Chair at That Microsoft Engineer Who Left for Google." *Business Insider,* July 26, 2017. Retrieved from http://www.businessinsider.com/steve-ballmer-i-didnt-throw-the-chair-at-th-google-engineer-2017-7.

McCulloch, W. S. "Why the Mind Is in the Head." In *Embodiments of Mind* (2nd ed.), 86–87. Cambridge, MA: MIT Press, 1951/1988.

Moriarty, B. *Trinity.* Cambridge, MA: Infocom, 1986.

Morrow, R. L., Garland, E. J., Wright, J. M., MacLure, M., Taylor, S., & Dormuth, C. R. "Influence of Relative Age on Diagnosis and Treatment of Attention-Deficit/Hyperactivity Disorder in Children." *Canadian Medical Association Journal* 184, no. 7 (April 17, 2012): 755–62. https://doi.org/10.1503/cmaj.111619.

Musil, R. *The Man Without Qualities.* Translated by B. Pike & S. Wilkins New York: Vintage International, 1995.

Naur, P. *Computing: A Human Activity.* November 15, 1990. Retrieved from http://www.naur.com/comp/p.html.

Nelson, G. *The Inform Designer's Manual* (4th ed.). St. Charles, IL: The Interactive Fiction Library, 2001. Retrieved from http://inform-fiction.org/manual/DM4.pdf.

Nettle, D. *Personality: What Makes You the Way You Are.* Oxford: Oxford University Press, 2007.

Open Science Collaboration. "Estimating the Reproducibility of Psychological Science." *Science* 349, no. 6251 (2015): 943. https://doi.org/10.1126/science.aac4716.

Pagels, H. *The Dreams of Reason.* New York: Bantam, 1988.

Pang, A. S.-K. "The Xerox PARC Visit." Making the Macintosh, n.d. Retrieved from https://web.stanford.edu/dept/SUL/sites/mac/parc.html.

Papert, S. *Mindstorms.* New York: Basic Books, 1980.

Pasquale, F. *The Black Box Society.* Cambridge, MA: Harvard University Press, 2015.

Patterson, M. L. "Okay, Feminism, It's Time We Had a Talk About Empathy." *Medium,* October 14, 2013. Retrieved from https://medium.com/@maradydd/okay-feminism-its-time-we-had-a-talk-about-empathy-bd6321c66b37.

Peirce, C. S. "Questions Concerning Certain Faculties Claimed for Man." *Journal of Speculative Philosophy* 2 (1868): 103–14.

Perlis, A. "Epigrams on Programming." *SIGPLAN Notices* 17, no. 9 (1982): 7–13.

Peterson, J. *Playing at the World.* San Diego, CA: Unreason Press, 2012.

———. "The First Female Gamers." *Medium,* October 5, 2014. Retrieved from https://medium.com/@increment/the-first-female-gamers-c784fbe3ff37.

Pike, R. "Notes on Programming in C." February 21, 1989. Retrieved from http://doc.cat-v.org/bell_labs/pikestyle.

Poincaré, H. *The Value of Science.* Translated by G. B. Halstead. New York: The Science Press, 1907.

Pólya, G. *How to Solve It: A New Aspect of Mathematical Method* (2nd ed.). Princeton, NJ: Princeton University Press, 1957.

Ramsey, F. P. "The Foundations of Mathematics." In *The Foundations of Mathematics and Other Logical Essays.* London: Routledge & Kegan Paul, 1931.

Reddy, V. *How Infants Know Minds.* Cambridge, MA: Harvard University Press, 2008.

Regier, D. A., Narrow, W. E., Kuhl, E. A., & Kupfer, D. J. "The Conceptual Development of *DSM-V.*" *American Journal of Psychiatry* 166, no. 6 (June 1, 2009): 645–50. Retrieved from https://doi.org/10.1176/appi.ajp.2009.09020279.

Reichenbach, H. *Experience and Prediction: An Analysis of the Foundations and the Structure of Knowledge.* Chicago: University of Chicago Press, 1938.

Roederer, J. *Information and Its Role in Nature.* Berlin: Springer, 2005.

Rosenoer, J. *CyberLaw: The Law of the Internet.* New York: Springer, 1997.

Rossiter, M. W. *Women Scientists in America: Forging a New World Since 1972.* Baltimore, MD: Johns Hopkins University Press, 2012.

Rota, G.-C. *Indiscrete Thoughts.* Boston: Birkhäuser, 1997/2008.

Sankis. "Dwarf Fortress — Boatmurdered." Let's Play Archive, 2007. Retrieved from http://lparchive.org/Dwarf-Fortress-Boatmurdered/Update%201-27/.

Sapolsky, R. M. *Behave: The Biology of Humans at Our Best and Worst.* New York: Penguin, 2017.

Schatzberg, A. F., Scully, J. H. Jr., Kupfer, D. J., & Regier, D. A. "Setting the Record Straight: A Response to Frances Commentary on *DSM-V*." *Psychiatric Times,* July 1, 2009. Retrieved from http://www.psychiatrictimes.com/dsm-5-0/setting-record-straight-response-frances-commentary-dsm-v.

Shanon, B. *The Antipodes of the Mind: Charting the Phenomenology of the Ayahuasca Experience.* Oxford: Oxford University Press, 2003.

Shapiro, F., & Forrest, M. S. *EMDR: The Breakthrough Therapy for Overcoming Anxiety, Stress, and Trauma.* New York: Basic, 2016.

Shepherd, E. "Some Assembly Required: Incremental Progress." n.d. Retrieved from ftp://ftp.apple.asimov.net/pub/apple_II/documentation/programming/6502assembly/Some%20Assembly%20Required%20Shepherd.txt.

Shorter, E. *What Psychiatry Left Out of the DSM-5.* New York: Routledge, 2015.

Singer, N. "Mapping, and Sharing, the Consumer Genome." *New York Times.* Retrieved from http://www.nytimes.com/2012/06/17/technology/acxiom-the-quiet-giant-of-consumer-database-marketing.html.

———. "Acxiom Lets Consumers See Data It Collects." *New York Times,* August 31, 2013. http://www.nytimes.com/2013/09/01/business/a-data-broker-offers-a-peek-behind-the-curtain.html.

Smullyan, R. *What Is the Name of This Book?* Englewood Cliffs, NJ: Prentice-Hall, 1978.

Spielman, R. M. *Psychology.* Houston, TX: OpenStax, 2016. Retrieved from Boundless Psychology.

StarkRavingMad. "Dwarf Fortress—Boatmurdered." Let's Play Archive, 2007. Retrieved from http://lparchive.org/Dwarf-Fortress-Boatmurdered/Update%201-30/.

Sullivan, D. "Racist Listings in Google Maps That Will Shock You & Why They May Be Happening." *Marketing Land,* May 20, 2015. Retrieved from http://marketingland.com/racist-listings-in-google-maps-129447.

"The Digint Programme." *The Intercept,* June 7, 2016. Retrieved from https://theintercept.com/document/2016/06/07/digint-narrative/.

Torvalds, L. "Re: Licensing and the Library Version of Git." Lwn.net, July 27, 2006. Retrieved from http://lwn.net/Articles/193245/.

Toyama, K. *Geek Heresy: Rescuing Social Change from the Cult of Technology.* New York: PublicAffairs, 2015.

"Trending Review Guidelines." Facebook Newsroom, May 2016. Retrieved from https://fbnewsroomus.files.wordpress.com/2016/05/full-trending-review-guidelines.pdf.

Trevarthen, C. "Making Sense of Infants Making Sense." *Intellectica* 34 (2002): 161–88.

Truesdell, C. *An Idiot's Fugitive Essays on Science: Methods, Criticism, Training, Circumstances* (2nd ed.). New York: Springer, 1984.

Turow, J. *The Daily You: How the New Advertising Industry Is Defining Your Identity and Your Worth.* New Haven, CT: Yale University Press, 2012.

Vygotsky, L. *Thought and Language.* Edited by A. Kozulin. Cambridge, MA: MIT Press, 1986.

———. *The Collected Works of L. S. Vygotsky: Problems of General Psychology.* Edited by R. W. Rieber & A. S. Carton. Translated by N. Minick. New York: Plenum, 1987.

"Well-Known Rationals." Keirsey.com, n.d. Retrieved from http://www.keirsey.com/4temps/famous_rationals.asp.

Williams, J. *Stoner.* New York: Viking Press, 1965.

Williams, P., & Tribe, A. *Buddhist Thought.* London: Routledge, 2000.

Winterhalter, B. "ISTJ? ENFP? Careers hinge on a dubious personality test." *Boston Globe,* August 31, 2014. Retrieved from http://www.bostonglobe.com/opinion/2014/08/30/istj-enfp-careers-hinge-dubious-personality-test/8ptUGXhu6DndFdjCngcxSN/story.html.

Wittgenstein, L. *The Blue and Brown Books* (2nd ed.). New York: Harper & Row, 1960.

Woods, M. "wood s lot." July 13, 2016. Retrieved from http://web.ncf.ca/ek867/wood_s_lot.html.

World Health Organization. "Gender." August 2015. Retrieved from http://www.who.int/mediacentre/factsheets/fs403/en/.

Wu, X., & Zhang, X. "Automated Inference on Criminality Using Face Images." November 21, 2016. Retrieved from https://arxiv.org/pdf/1611.04135v2.pdf.

Zolli, A. "Darwin's Stickers." *Radiolab,* February 9, 2015. Retrieved from http://www.radiolab.org/story/darwins-stickers/.

INDEX

Page numbers in *italics* refer to illustrations.

ILLUSTRATION CREDITS

Page iii: Courtesy of Richard McGuire
Page 21: By permission of Mark Simonsen
Page 28: Courtesy of Brecht Evens, from "The City of Belgium," *Drawn & Quarterly,* 2018
Page 125: Illustration from *Epileptic,* by David B., copyright © 2005 by L'Association, Paris, France. Used by permission of Pantheon Books, an imprint of the Knopf Doubleday Publishing Group, a division of Penguin Random House LLC. All rights reserved.
Page 141: Courtesy of Craig VanGrasstek
Page 152: Courtesy of Tim Denee
Page 162: Courtesy of Laurence Jenkins
Page 166: Courtesy of Knee Jerk
Page 252: Courtesy of Matt Jones
Page 253: Courtesy of Matt Jones

A NOTE ABOUT THE AUTHOR

David Auerbach is a writer and software engineer who has worked for Google and Microsoft. His writing has appeared in *The Times Literary Supplement, MIT Technology Review, The Nation, The Daily Beast, n+1,* and *Bookforum,* among many other publications. He has lectured around the world on technology, literature, philosophy, and stupidity. He lives in New York City.

A NOTE ON THE TYPE

This book was set in Minion, a typeface produced by the Adobe Corporation specifically for the Macintosh personal computer, and released in 1990. Designed by Robert Slimbach, Minion combines the classic characteristics of old-style faces with the full complement of weights required for modern typesetting.

Composed by North Market Street Graphics, Lancaster, Pennsylvania

Printed and bound by LSC Communications, Harrisonburg, Virginia

Designed by Maggie Hinders